MARITIME SAFETY

MARITIME SAFETY

The Human Factors

Dr Sean M Trafford

Book Guild Publishing
Sussex, England

First published in Great Britain in 2009 by
The Book Guild Ltd
Pavilion View
19 New Road
Brighton, BN1 1UF

Typesetting in Times by
YHT Ltd, London

Printed in Great Britain by
CPI Antony Rowe

A catalogue record for this book is available from
The British Library.

ISBN 978 1 84624 379 0

This book is dedicated to all members of the international seafaring community who deserve the utmost admiration for the work they do, frequently under challenging conditions and often under difficult circumstances.

Contents

PART III CONCLUSION

List of Figures and Tables

Foreword

As a result of rising maritime accident rates and pollution incidents in the 1970s and 1980s, the International Maritime Organisation introduced two internationally recognised Codes that entered fully into force in 2002: the International Management Code for the Safe Operation of Ships and for Pollution Prevention (ISM Code), and the 1995 revision of the 1978 Convention on Standards of Training, Certification and Watchkeeping for Seafarers (STCW Code).

Introduction of the Codes served to focus attention on the need to raise maritime safety standards world-wide and a great deal of literature has since been devoted to the practicalities of implementing the provisions of the two Codes. However, little consideration has been given to whether or not it is in fact possible to develop a safety culture across a fragmented, global industry and to what extent the global diversity of cultures might impact upon attempts to implement common standards of safety throughout such an industry. This book addresses those particular areas of concern by examining the human factors involved.

The book should not be approached as a work of academic reference wherein all academic arguments relating to behavioural safety may be found, nor on the other hand is it intended to be merely an introductory textbook to behavioural safety in a maritime context. Rather, it should be regarded as a practitioner's handbook, a text of interest to those active in the industry world-wide such as ship managers and port operating companies as well as safety professionals charged with maintaining company safety management systems.

However, the book does also have an academic aspect, based as it is upon an earlier academic research study by the author using a rigorous methodology complemented by a detailed comparative case study. Indeed, the book has been written with the intention that in addition to being of interest to maritime professionals it may be useful to universities offering postgraduate degrees in maritime operations and management, as well as to technical colleges and ship training institutes providing courses in nautical studies, shipboard management and ship operations.

In Part I of the book the history of maritime safety legislation is initially summarised in order to provide a contextual setting for the subsequent review of salient literature relating to safety management, the nature of safety as a concept and the source of the legal and moral authority of safety regulations.

Literature relating to culture and its effects upon human behaviour is also examined in order to facilitate identification of socio-cultural factors that impact upon the development of safety management systems, the implementation of safety procedures and the establishment of effective training policies in individual shipping organisations. Those constraints and pressures, together with others associated with both national and corporate economic realities and the psychological dispositions of individuals, are then superimposed upon a structural model of the ISM Code to produce a dynamic model that provides a predictive and investigative tool for use by managers and consultants charged with improving an organisation's safety performance.

Part II of the book utilises case studies to illustrate the practical effects of the socio-cultural, psychological and economic constraints and pressures identified in the first part of the book and why good employment policies are important, not only for a company trying to implement good safety practices but also for the development of a safety culture throughout the wider industry. The case studies are based upon actual research undertaken by the author in two shipping companies with similar organisational structures but diametrically opposed economic and cultural parameters.

Finally, in Part III of the book a review is carried out of the general themes and salient points developed in the previous sections. Based upon this review the book concludes by addressing the vexed question of how to achieve a common standard of safety throughout the shipping industry: whether the better path to follow is to ensure rigorous enforcement of common rules and regulations by State Administrations or whether an industry-wide safety culture can be more readily established by embracing self-regulation and placing more emphasis on ensuring the development of suitable education and training programmes of a common standard and implementing them across a spectrum of individuals having diverse cultural norms.

PART I

MANAGING MARITIME SAFETY:
THEORY AND CONCEPTS

1

Background and Overview

Getting to Know the Ropes

This opening chapter provides an overview of the development and implementation of maritime safety regulations and how the shipping industry has attempted to improve safety standards despite the effects of industry fragmentation, cultural diversity and an international division of labour.

INTRODUCTION

The overriding statutory instrument governing operational safety within the shipping industry today is the International Management Code for the Safe Operation of Ships and for Pollution Prevention, otherwise known as the International Safety Management Code or simply the ISM Code. A great deal has been written about the efficacy of the ISM Code, whether or not it is 'working' and why some ship owners and some flag states appear to intentionally avoid implementing the provisions of the Code, some writers suggesting that there are economic advantages to be gained by avoiding its implementation. A great deal has also been written about how the provisions of the Code are to be interpreted and what they actually require ship owners to do in order to comply with the Code.

However, the fact that the ISM Code is itself a relatively brief and concise document comprising only some 39 pages inclusive of annexes and guidelines on implementation, gives one cause to reflect upon why so much has been written about it and why so much debate has surrounded its introduction. At a time when safety has become of paramount importance in all areas of public life and in private industry, could it be that people in general have little

3

understanding of the fundamental tenets of safety, from where safety legislation and safety procedures draw their authority, whether people of different cultures view safety from the same perspective and whether an individual's psychological disposition may affect his or her judgement when making decisions that involve risk taking?

It is frequently asserted that approximately 80% of all shipping accidents are attributable to human error, but according to Lord Donaldson,[1] 'Virtually all accidents, marine or otherwise, are indeed caused by human error, but the real issue is what was the precise nature of the error and when did it occur? ... It may have been historic or it may be current. In a marine context it may have occurred at the stage of design, construction, operation, supervision or whatever. Unless you answer these questions and a good few more, you are unlikely to know what to do to improve safety.'

This book sets out to provide a better understanding of the factors that may lead to human error, first by examining the intrinsic precepts of safety and then identifying and examining the obstacles that the global diversity of cultures, individual psychological dispositions, economic pressures and socio-economic conditions present to the development and implementation of safe practices throughout the shipping industry and by what means it is possible to overcome those obstacles.

Using material drawn from a review of the relevant literature and the provisions of the ISM Code itself, a multi-stage model of the working of the ISM Code is developed, in part to assist with making predictions regarding the effectiveness with which safety management systems of individual companies will meet the objectives of the Code and in the main to provide an organising framework to help identify:

- Where and in what manner cultural, psychological and economic pressures might reasonably be expected to impact upon the interpretation and implementation of the provisions of the ISM Code.
- At which particular levels of safety management interventions are necessary to deal with the effects of those influences on the implementation of the provisions of the ISM Code in shipping organisations.

This book also addresses the question of whether stricter enforcement of existing regulations, possibly by Port State Control inspectorates, a practice that recognises heterogeneity in the shipping industry, or greater emphasis on education and training such as that provided for in the 1995 revision of the 1978 Convention on Standards of Training, Certification and Watchkeeping for Seafarers (STCW Code), an attempt at introducing homogeneity to the shipping industry, is the better path to follow to counter the effects of those culturally influenced constraints and pressures upon the way in which shipping companies interpret and implement the ISM Code.

BACKGROUND TO SAFETY LEGISLATION IN THE SHIPPING INDUSTRY

Shipping is inherently an industry that operates within a global context. A ship built in one country may be owned by a company incorporated in another country, flagged out to a third country, managed by a company in another country, crewed by nationals of yet another country and trade internationally. This has led to difficulties both in establishing and in policing common safety standards throughout the industry. However, if safety rules differ or are interpreted differently from one country to another, then an unnecessary and unwanted element of uncertainty and possibly confusion would be introduced into international maritime voyages with the potential for serious resultant consequences.

Ninety-five percent of world trade by volume – raw materials, finished goods and energy supplies – is transported by sea[2] and a large amount of capital is invested in shipping. The world-wide shipbuilding order book stood at US$54 billion in June 2001, with a total of 1824 new vessels due for delivery within the following two to three years.[3] By July 2005 the number of ships on order had increased to 4324 vessels totalling 226.7 million dwt.[4] Internationally recognised standards of safety within the shipping industry are therefore, of the utmost importance, not only to ensure human safety and protection of the marine environment, but also for purely economic reasons.

In order to achieve an industry-wide safety culture there must be rules and regulations specifying minimum standards for the building, operation and maintenance of ships, and there must also be a common understanding of those rules and regulations. Procedural and technical safety rules and regulations result from past experience, technical progress, scientific research, reflective deliberation and risk assessment and, where necessary, they are codified by state legislatures. The resultant statutes, such as the Health and Safety at Work etc Act 1974 in England and Wales for example, are implemented by public and commercial organisations, monitored by executive bodies such as the Health and Safety Executive in the United Kingdom and, in the event of severe breaches, enforced by the judiciary.

Many safety rules also exist that are not subject to legislative enforcement but do carry persuasive force. The Rules for the Classification of Steel Ships developed by the classification societies such as Bureau Veritas and Lloyds Register of Shipping, the standards developed by the Society of International Gas Tanker and Terminal Operators (SIGTTO) and the standards developed by the Oil Companies International Marine Forum (OCIMF) are examples of safety rules that have persuasive rather than legislative force.

Seafaring has traditionally been regarded as a hazardous occupation. Historically, the main hazards were associated with perils of the sea because sailing vessels were very much at the mercy of the elements and safety training comprised mentoring of new entrants to the industry by experienced seafarers. Safety practices were often summed up in epithets such as 'one hand for the ship and one for yourself' and 'get to know the ropes'. However, not all hazards were the result of natural providence: poor workmanship, overloading of vessels and failure to maintain vessels in a seaworthy condition also played their part.

With the advent of the industrial revolution came the use of iron and steel to construct ships and development of steam and internal combustion engines to power them. Ships were no longer so susceptible to the vagaries of the elements. The invention of the telegraph in the mid nineteenth century made rapid communications possible, and by the latter part of the nineteenth century a global communications network was in existence. Ever more rapid and ever cheaper means of communication, coupled with ships that were no longer dependent upon wind power for propulsion, resulted in more companies being able to trade in a much wider marketplace than ever before.

With the industrial revolution at its height in Britain there was an increasing need for imports to supply raw materials for the fabrication of finished goods for export. International trade soared. As Amin[5] quoting Hirst and Thompson pointed out, at the height of the imperial age between 1878 and 1914 international flows of investment, exports and people exceeded current levels.

However, commensurate with technical progress and the increase in shipping in the nineteenth century, there was also an increase in the losses of ships and their crews. New hazards more readily associated with industrial accidents than with perils of the sea were introduced to seafaring: boiler explosions, machinery failures, injuries and deaths due to unsafe operation of machinery, heat exhaustion and asbestosis to name but a few. In the social climate prevailing at that time industrial injuries were generally accepted as occupational hazards and the impetus to reduce accident rates came predominately from social reformers and, perhaps more frequently, from industrialists seeking to reduce the costs engendered as a result of damage to property, equipment and machinery.

But there was also growing public concern about maritime safety and in 1836 in response to that concern the British government appointed a committee to investigate the growing number of shipwrecks. Subsequently, in 1850 legislation was passed creating the Marine Department of the Board of Trade with a brief to enforce laws governing crewing, crew competency, and the operation of merchant vessels. Subsequently regulations were introduced requiring masters, mates and engineers of British ships to be in possession of Certificates of Competency, issued by the Board of Trade. Seafarers could sit examinations for the Certificates in centres worldwide and the Marine Department of the Board of Trade was responsible for ensuring uniformity of standards at all examination centres.

Then in 1914 the British government convened an international conference with the aim of establishing international maritime safety regulations. The outcome of the conference was the adoption of the first Convention for the Safety of Life at Sea (SOLAS). According to Özçayir[6] the title of the Convention was significant because it was the first time in shipping history that protection of human life rather than vessels and their cargoes became a priority. However, due to the outbreak of the First World War the SOLAS Convention of 1914 was not implemented. It was not until 1929 when the United Kingdom hosted another conference that a second SOLAS

Convention was adopted, the Convention entering into force in 1933. Subsequently, in 1948 at a further conference hosted by the United Kingdom the third SOLAS Convention was adopted and in the same year the newly established United Nations Organisation began to take an interest in the regulation of international shipping.

ESTABLISHMENT OF THE INTERNATIONAL MARITIME ORGANISATION

From the mid-nineteenth century to the mid-twentieth century the United Kingdom dominated the international shipping industry. British shipping companies carried over 50% of world trade and owned a similar percentage of world tonnage, and Britain was the world's leading shipbuilding nation.[7] This enabled the United Kingdom government to exercise sufficiently strong administrative control to impose uniform, internationally recognised maritime rules throughout the world shipping industry.

An example of this was the introduction of the load line, collo-quially known as the Plimsoll mark after Samuel Plimsoll, the par-liamentarian who championed its introduction in order to stop the dangerous practice of overloading vessels. In response to public pressure to improve the safety of shipping, the British government enacted the Merchant Shipping Act of 1875 providing *inter alia* for the marking of a load line on the hull of every ocean-going cargo ship flying the British flag or leaving a British port. Since the United Kingdom not only dominated world shipping but also had a large empire at that time, this regulation applied to the majority of the world's ships, and similar regulations were subsequently adopted by other states for vessels flying their own national flags resulting in the *de facto* introduction of the International Load Line.

Political and sociological changes during the twentieth century saw the decline of Britain's domination of international shipping and a consequent increase in organisational fragmentation and cultural diversity within the industry. Today, ship ownership is pre-dominately in the hands of Greek and Chinese companies,[8] the leading ship-building nations are Japan and Korea,[9] and ships' crews are drawn mainly from the former eastern bloc countries, the Phi-lippines and the Indian sub-continent[10] with Filipino seafarers accounting for approximately 20% of the shipping industry's total labour force.[11]

This administrative, organisational and cultural fragmentation within the shipping industry is the manifestation of a phenomenon recognised by Fröbel et al[12] as a new international division of labour. According to Fröbel, global reproduction of Western economic institutions has resulted in the transplantation of many Western dominated industries to other regions around the world where cheaper labour forces are available, resulting in considerable restructuring of trans-national companies and industries, and subsequent problems associated with that restructuring.

Fragmentation in the shipping industry in the twentieth century led to increasing difficulties in establishing and policing uniform safety standards throughout the industry. Consequently, in 1948 the United Nations Organisation (UN) convened a conference in Geneva to consider regulation of international shipping, which led to the introduction of a Convention establishing the International Maritime Consultative Organisation (IMCO) as an agency of the United Nations. The Convention was ratified in 1958 by a quorum of UN member states and IMCO, renamed the International Maritime Organisation (IMO) in 1982, convened its first meeting in 1959. Today IMO has its headquarters in London and is recognised as the international regulatory body for world shipping.

IMO, SOLAS AND ECONOMIC FORCES

When a qualified majority of IMO member states ratify an IMO Convention the provisions of that Convention enter into force. By this means IMO introduced a series of internationally applicable regulations, commencing with the 1960 revision of the 1948 SOLAS Convention, followed inter alia by the 1974 Convention on Safety of Life at Sea (SOLAS) with subsequent amendments and the 1978 Convention on Standards of Training, Certification and Watchkeeping for Seafarers (STCW Convention) which was substantially revised in 1995.

But according to a report in 1996 by the Organisation for Economic Cooperation and Development (OECD)[13] the legislative approach had not been effective in raising levels of safety in shipping operations and practice, possibly because the resources necessary to police the situation have not always been forthcoming and also there are economic advantages to be gained by ship owners who avoid following the rules. The report concluded that enforcement and implementation of SOLAS regulations varied from one country to another and from one owner to another, with unscrupulous owners disregarding even the most basic safety requirements.

Contrary to the view expressed in the OECD report, the United Kingdom Health and Safety Executive (HSE)[14] is of the opinion that short-term financial gains resulting from perceived competitive advantages from non-observance of rules and standards agreed by convention to be safe, might well be outweighed by long-term economic disadvantages, the appreciation of which might in itself be subject to differing cultural interpretation between companies based in developing countries and those based in developed countries.

The impact of economic forces in today's shipping industry may also have been somewhat mitigated by the current profitability of the industry. Beale[15] estimated the 2003 earnings of ship owners at US$110bn and their profit at $80bn resulting from the 'best-ever trading conditions'. Beale also noted that despite the world fleet recently increasing in size and value, ship owners were paying 30% less in insurance premiums than a decade before. The presumption therefore is that ship owners are getting much better returns on their investments than hitherto and economic forces are thus less pressing than they were previously.

The significance of economic forces on the way a business is managed was also questioned by Drucker[16] who argued that people, not forces, create and manage a business and management's actions are not determined, but merely constrained, by economic forces. Constraints imposed by economic forces are, however, generally recognised as being important influences on the development and implementation of safety management systems, safe working practices and personal safety awareness.[17]

THE INTERNATIONAL SAFETY MANAGEMENT CODE

In 1994, in an endeavour to ensure uniform implementation of the provisions of international Conventions regarding maritime safety and environmental protection, IMO introduced, as an amendment to the 1974 SOLAS Convention, the International Management Code for the Safe Operation of Ships and for Pollution Prevention, otherwise known simply as the ISM Code, which applies not only to ships and their crews but also to ship owners, managers and operating companies. In the terms used by Lau *et al*[18] drawing on McMahan and Woodman,[19] the introduction of the ISM Code was a standard system-wide intervention.

Central to the ISM Code is a requirement for each ship-operating company to establish and implement a safety management system (SMS). However, the ISM Code is prescriptive in outcome rather than prescriptive in process, and in recognition of the global nature of shipping and the wide differentials existing within the industry the Code was drafted in broad terms based upon general principles and objectives *(ISM Code, Preamble, para. 4 and 5)*, allowing sufficient latitude to accommodate the specific needs and varied conditions facing individual ship owners and operators.

The 1996 OECD report, although written prior to implementation of the ISM Code, acknowledged that its impending introduction represented the latest attempt to improve minimum standards of safety and provided owners and operators with a useful industry-wide organisational framework to coordinate the improvement effort.

Implementation of the ISM Code was predicted to result in lower insurance claims and hence lower insurance premiums.[20] The prediction was supported by a survey carried out in 2001 by the Swedish Club, a ship owners' mutual protection and indemnity insurance association, which indicated that members of the club who had fully implemented the ISM Code provisions had seen reductions of more than 30% in hull insurance claims and a similar improvement in protection and indemnity insurance claims. The findings support the HSE argument that any short-term economic advantages gained by non-observance of safety regulations are outweighed by long-term advantages gained by observance of the rules.

However, there remains a great deal of debate within the shipping industry concerning the efficacy of the ISM Code.[21] Some writers believe that the Code is being ignored by ships' senior officers for fear that in the event of an accident and subsequent investigation data recorded under the provisions of an approved SMS may be used against them or their employers in a court of law.[22] Other writers report a general negative response from the industry as a whole with a growing number of key players writing off the industry's response to the ISM Code as a flop.[23]

THE STCW CODE

Introduction of the 1978 International Convention on Standards of Training, Certification and Watchkeeping for Seafarers (STCW) was an attempt to improve maritime safety by establishing internationally agreed minimum requirements for the training, certification and watchkeeping for seafarers. According to IMO,[24] 'Previously the standards of training, certification and watchkeeping of officers and ratings were established by individual governments, usually without reference to practices in other countries. As a result standards and procedures varied widely, even though shipping is the most international of all industries.'

The Convention, which prescribed the minimum standards for seafarers that countries were obliged to meet or exceed in relation to training, certification and watchkeeping, did not deal with manning levels since IMO provisions in this area are covered by a regulation in Chapter V of the International Convention for the Safety of Life at Sea (SOLAS), 1974, supported by resolution A.890(21) Principles of Safe Manning, adopted by the IMO Assembly in 1999, as amended by Resolution A.955(23) Amendments to the Principles of Safe Manning (Resolution A.890(21)).

Rapid technological advances in the second half of the twentieth century, smaller crews and faster port turnarounds significantly reduced the opportunities for on-board training and familiarisation of crews. This was of particular importance because of the increasing use of crews from developing nations and the fact that application of the Convention not only lacked detailed guidelines but also left to individual administrations the responsibility for ensuring that the Convention was satisfactorily implemented. This inevitably led to widespread differences in the way that individual administrations interpreted and enforced the provisions of the Convention and consequently a loss of confidence in certain national Certificates of Competency. It is fair to say that the 1978 Convention was never uniformly applied and did not impose any strict obligations on parties regarding implementation.

The STCW Convention of 1978 was therefore substantially

updated in 1995 by means of amendments that formed an Annex to the Convention. The amendments were in fact so extensive that effectively the 1978 Convention had been rewritten. As noted by IMO, 'The 1995 amendments, adopted by a Conference, represented a major revision of the 1978 STCW Convention, in response to a recognized need to bring the Convention up to date and to respond to critics who pointed out the many vague phrases, such as "to the satisfaction of the Administration", which resulted in different interpretations being made.' The 1995 amendments entered into force on 1st February 1997, although until 1st February 2002 parties could continue to issue, recognise and endorse Certificates issued before that date to seafarers who began training or seagoing service before 1st August 1998.

ADMINISTRATION OF THE ISM & STCW CODES

The fundamental principle embodied in Article 92(1) of the 1982 Law of the Sea Convention (UNCLOS) is that jurisdiction over a vessel on the high seas may be exercised only by the administration of the state whose flag the vessel is entitled to fly. There also exists a general presumption under customary international law that a flag state shall ensure vessels flying its flag implement the provisions of Conventions which the state has ratified and adopted.

But not all flag states have been equally conscientious in ensuring that vessels flying their national flag comply with the provisions either of the ISM Code or the STCW Code. As a result, the principle of sole jurisdiction by flag states has been eroded in recent years with *ad hoc* powers of inspection and detention of vessels being exercised by port states, i.e. those littoral states whose ports a vessel enters or through whose territorial waters a vessel sails.

One notable feature of the STCW Convention is that the Articles of the Convention include requirements relating to issues surrounding certification and port state control. In particular, the Convention applies to ships of non-party states when visiting ports of states which are parties to the Convention. Article X requires parties to apply the control measures to ships of all flags to the extent necessary to ensure that no more favourable treatment is given to ships entitled to fly the flag of a state which is not a party than is given to ships entitled to fly the flag of a state that is a party. The difficulties that could arise for ships of states which are not parties to the Convention may be one reason why the Convention has received such wide acceptance. By December 2000 the STCW Convention had 135 contracting parties, representing 97.53% of world shipping tonnage.

But if individual nation states differ in the degree to which they ensure that ships flying their flags comply with internationally agreed rules and regulations, possibly because the states have markedly different cultural values and norms from each other, then the enforcement by a port state of the obligations and duties accepted by a flag state as a party to an internationally recognised Convention can clearly lead to conflicts of opinion between the states in question regarding the seaworthiness of a vessel and the acceptability of its on-board safety procedures.

Furthermore, since individual safety perspectives are influenced by various factors some of which, such as education and training, may themselves be influenced by national culture, it is logical to suppose that national cultures might also impact upon how seagoing personnel conceptually perceive safety and how safety regulations are interpreted and implemented by people of different cultures in management positions ashore.

The apparent non-observance of the provisions of the SOLAS Convention by some ship owners may therefore owe more to cultural diversity within the shipping industry and the impact of national cultures upon the interpretation of the ISM and STCW Codes than to perceptions of short-term gain, fear of retribution, or mere negative attitude.

CULTURE AND SAFETY

The contextual environment in which the shipping industry operates is by its very nature culturally diverse, and therefore cross-cultural management plays an important role in the management and operations of shipping organisations, not only with regard to national and corporate cultures but also to safety cultures.

Trompenaars[25] quoting Geertz[26] defines national culture as the means by which people communicate, perpetuate and develop their knowledge about attitudes towards life. Culture is the fabric of

14

meaning in terms of which human beings interpret their experience and guide their action. Trompenaars continues by offering a means of measuring culture, arguing that cultures can be distinguished from each other by the differences in shared meanings they expect and attribute to their environment.

Schein,[27] on the other hand, implies that for organisational studies to advance, culture needs to be observed more than measured. The views of Trompenaars and Schein may not be mutually exclusive, however, since the former's approach may quantify the degree to which cultures differ and the latter's approach may indicate how those differences may impact on organisational functions such as management styles and risk assessment.

A number of value frameworks for use in international business research have been compiled, the most common being the dimensions identified by Hofstede[28] and Trompenaars along which national cultures differ from one another, each dimension having quite distinct characteristics. The cultural dimensions represented by the value frameworks provide the context within which transnational companies operate and present a knowledge management task to be dealt with by the application of cross-cultural management techniques.[29]

Trompenaars notes that culture manifests itself at different levels, such as national or regional cultures, corporate cultures and the culture of particular functions within organisations such as marketing, or research and development. Organisations also frequently refer to safety cultures, and the Advisory Committee on the Safety of Nuclear Installations (ACNSI)[30] commented in its third report that safety culture is a sub-set of, or at least profoundly influenced by, the overall culture of an organisation.

Some authors such as Cooper[31] and Young[32] argue that far from reflecting shared values and beliefs, corporate culture is the result of conflict and alignment of many sub-cultures within an organisation. According to this concept, corporate culture is a heterogeneous not a homogeneous phenomenon, and Cooper questions whether an industry-wide homogeneous safety culture can ever arise, let alone a global one.

Others such as Amin (cited earlier) and Spybey[33] propose that globalisation has blurred the differences between cultures. They hypothesise that people are influenced by their environment and that influence takes place partly through institutionalised patterns of social interaction which are, in a globalised society, global institutions.

It can also be argued that whilst the twin pillars of globalisation, increasingly rapid communications and increasingly rapid transportation, have brought about an ever-increasing rate of exposure of cultures to each other,[34] the effects of such exposure are selective, different cultures responding differently to the same external agency.

SAFETY, EDUCATION AND TRAINING

In addition to concluding that the legislative approach has not made much of an impact on raising the levels of safety in shipping operations and practice, the 1996 OECD report also expressed concerns about the professional training of seafarers. However, because the OECD report was published in 1996 its authors were not in a position to assess the effects of the 1995 revision of the 1978 Convention on Standards of Training, Certification and Watchkeeping for Seafarers undertaken by IMO in an endeavour to provide a common minimum standard of education and training for seafarers throughout the shipping industry.

The revised Convention, which entered fully into force in February 2002, strongly emphasises the need for greater safety training. But despite the notion that safety training will cure most ills in regard to accidents, evidence exists to show that it is not always effective,[35] which may be related to the variability of the quality of training given or to the cultural attitudes of the trainees. Furthermore, safety training should not be seen as a substitute for, or adjunct to, professional training but as an integral part of professional training.

Locus of control, a personality construct developed by Rotter[36] and subsequently extensively researched, is relevant in the context of education and training. The construct recognises that some people perceive the outcome of their actions as being controlled principally by themselves while others perceive the outcome of their actions as being controlled principally by external factors. The former are said to have an internal locus of control orientation and the latter an external locus of control orientation.

From his observations Rotter concluded that given identical conditions for learning, different people learn different things from the same lesson dependent upon their locus of control orientation, which sits well with Hofstede's observation that different people react differently to the same external stimuli. This suggests there may be a significant correlation between locus of control orientation and an individual's cultural background, which may have important implications with regard to using education and training as a means of developing common attitudes towards safety in a global industry.

According to Rotter's and Hofstede's observations, and as confirmed by Hale, safety training even if integrated with professional training, may not be effective in establishing common safety standards within the shipping industry. If that is the case, then it would be better to recognise the existence of heterogeneous attitudes towards safety and to overcome such diversity by placing greater emphasis on policing and enforcing existing regulations to ensure individual vessels comply with the strictures of the ISM Code and other internationally agreed Conventions.

Holden, however, contends that cultural differences are not a barrier to trans-national organisational harmonisation but simply present another knowledge management task. He also argues that while culture shapes behaviour and influences one's view of the world, culture is also learned: it is therefore not a static dimension and can be changed. This implies that there may be a correlation between locus of control and an individual's cultural background, and also that there may be a chronological aspect to locus of control orientation such that it may shift from external to internal as people feel increasingly competent to control events in their lives, as they gain knowledge from experience, education and training.

If this were the case, then greater emphasis on education and training such as that provided for in the 1995 STCW Convention would be the better path to follow to achieve a more homogeneous attitude to safety among seafarers and the maritime community in general, thus ensuring that the objectives of the ISM Code are fulfilled.

SUMMARY

This chapter introduced the salient factors relevant to maritime safety that are expanded upon in subsequent chapters. The reader was invited to consider the need for common standards of safety throughout the shipping industry and the problems associated with implementing those standards world-wide. Historical factors that influenced the way in which maritime safety administration developed were outlined, from the decline of Britain's domination of international shipping, through the subsequent increase in organisational fragmentation and cultural diversity within the industry, to the creation of the International Maritime Organisation as an agency of the United Nations charged with responsibility for international maritime safety.

The reader was introduced to the SOLAS Convention, the provisions of which provided for the first time a framework for the safety not only of ships and their cargoes but also of personnel. It was explained how non-observance of the SOLAS Convention by some ship owners and some flag state administrations led to the subsequent introduction of the STCW and ISM Codes in an attempt to foster a genuine safety culture throughout the shipping industry.

Finally, the need for common standards of education and training to ensure a common standard of safety across a fragmented, culturally diverse industry operating in a global context was touched upon.

2

The ISM and STCW Codes

Standing Orders

ISM CODE OBJECTIVES

The ISM Code is a concise document that is prescriptive in outcome rather than in process. It is the objectives of the Code that are of the highest importance rather than specific procedures to be employed to achieve those objectives.

Clause 1 of the ISM Code Preamble states that the purpose of the Code is to provide an international standard for the safe management and operation of ships and for pollution prevention. Clause 1.2 of the Code refines this overall purpose by identifying three specific objectives, namely to ensure:

- Safety at sea,
- Prevention of human injury or loss of life, and
- Avoidance of damage to the marine environment.

The remainder of the Code is devoted to guidance on how the objectives are to be achieved, in particular the development and implementation of a safety management system (SMS) by each ship-operating company. Although the Code contains no specific procedures it does introduce generic functional requirements that are required to ensure the effectiveness of an SMS.

STCW CODE OBJECTIVES

In contrast to the ISM Code, the STCW Code is prescriptive in process rather than in outcome. The document is quite explicit and outlines in some detail what flag state administrations, companies and seafarers must do in order to comply with the provisions of the Code.

However, the STCW Code does have similar objectives to the ISM Code and the simplest way to illustrate those objectives is to quote from Attachment 2 to the Final Act of the Conference contained in the Annex to the STCW Convention[1] wherein the conference adopts the 1995 amendments to the Annex to the International Standards on Training, Certification and Watchkeeping of Seafarers (STCW) 1978:

> **RECOGNIZING** the importance of establishing detailed mandatory standards of competence and other mandatory provisions necessary to ensure that all seafarers shall be properly educated and trained, adequately experienced, skilled and competent to perform their duties in a manner which provides for the safety of life and property at sea and protection of the marine environment,
>
> **ALSO RECOGNIZING** the need to allow for the timely amendment of such mandatory standards and provisions in order to effectively respond to changes in technology, operations, practices and procedures used on board ships,
>
> **RECALLING** that a large percentage of maritime casualties and pollution incidents are caused by human error,
>
> **APPRECIATING** that one effective means of reducing the risks associated with human error in the operation of seagoing ships is to ensure that the highest practicable standards of training, certification and competence are maintained in respect of the seafarers who are employed on such ships,
>
> **DESIRING** to achieve and maintain the highest practicable standards for the safety of life and property at sea and in port and for the protection of the environment.

AN INTERNATIONAL STANDARD

The provision of an international standard as cited in the ISM Code Preamble implies that certain criteria are agreed by all IMO member states that are contracting states to the ISM Code Convention. This would involve considerable cross-cultural agreement since IMO has 166 member states of which 156 states with approximately 99% of the world's merchant fleet gross tonnage on their registers are contracting states to the Convention.[2]

Under the 'Guidelines on the Implementation of the ISM Code by Administrations' (the Guidelines) adopted under Resolution A.788(19), it is the flag state administration – i.e. the government of the state whose flag a ship is entitled to fly – that is responsible for verifying compliance with the ISM Code. By extension therefore, there is a presumption that all contracting states to the ISM Code have a common understanding of the aims and objectives of the Code.

This presumption, however, is not irrebuttable since the Preamble declares that the ISM Code is based on general principles and objectives and expressed in broad terms so that it can have widespread application. The inference therefore is that there is wide scope for interpretation of the provisions of the Code.

Indeed, the Guidelines at Clause 2.1.3 recommend that in determining conformity or non-conformity of the SMS elements specified in the ISM Code, administrations should limit the development of prescriptive management system solutions and leave the shipping company itself to develop the solutions which best suit that particular company, that particular operation or that particular ship.

Furthermore, the Guidelines at Clause 2.1.4 recommend that administrations should ensure that these assessments are based on determining the effectiveness of the SMS in meeting specified objectives rather than detailed requirements in addition to those contained in the ISM Code.

From this it is clear that it is the objectives specified in the Code that are of overriding consideration and companies are free to develop SMS elements in accordance with their operating requirements, provided that the SMS achieves the specific objectives of safety at sea, prevention of human injury or loss of life, and avoidance of damage to the marine environment, as outlined in the Code.

To remove some of the subjectivity and latitude afforded to companies developing an SMS, Clause 1.4 of the Code introduces functional requirements for safety management systems and elaborates on each requirement in further clauses. The correlation between the functional requirements and the elaborating clauses is shown in Table 1 below, adapted from the tabulation by Sagen.[3]

However, although providing guidance to shipping companies and

TABLE 1. SMS functional requirements[3]

Basic functional requirements	Elaboration clauses
1.4 Every company should develop, implement and maintain a safety-management system (SMS) which includes the following functional requirements:	Cl. 11
.1 A safety and environmental protection policy;	Cl. 2 / 1.2
.2 Instructions and procedures to ensure safe operation of ships and protection of the environment in compliance with relevant international and flag State requirements;	Cl. 6, 7 and 10
.3 Defined levels of authority and lines of communication between, and amongst, shore and shipboard personnel;	Cl. 3 and 5
.4 Procedures for reporting accidents and non-conformities with the provisions of this Code;	Cl. 9
.5 Procedures to prepare for and respond to emergency situations; and	Cl. 8
.6 Procedures for internal audits and management reviews	Cl. 12

administrations regarding the minimum functional requirements that must be included in an acceptable SMS, neither the functional requirements nor the elaboration clauses are prescriptive in nature. Thus, companies and administrations are still left with considerable latitude in how they address the requirements of the Code. There remains also a need for them to exercise subjective judgement both in interpreting the objectives of the Code and in deciding whether or not any particular SMS will lead to the objectives being satisfactorily addressed.

Unlike the ISM Code, the STCW Code does not assume that certain criteria are agreed by all IMO member states that are contracting parties to the STCW Convention, even though the stated objectives of the Convention are to introduce common minimum standards of education and training. Instead, the STCW Code details precisely what is expected of each flag state, shipping company and individual to ensure compliance with the Code, although as noted later in this chapter the Code's provisions may still be interpreted differently by people with dissimilar cultural backgrounds.

SAFETY CULTURE AND THE ISM CODE OBJECTIVES

Sagen terms the introduction of the ISM Code a paradigm shift in international ship operation and holds that the key factor in fulfilling the intentions and objectives of the ISM Code is the establishment of a safety culture in ship operation, noting the often quoted opinion of former IMO General Secretary William O'Neill that the ISM Code will provide ship owners with real business advantages, provided they truly want to change towards a safety culture.

By prefacing the words safety culture with the indefinite article rather than the definite article both Sagen and O'Neill are presumably using the term generically to imply the observance of internationally agreed rules and regulations and the introduction of industry-wide working practices and attitudes agreed by convention to be safe, not simply that there is only one safety culture acceptable throughout the entire shipping industry. Indeed, the former is a basic assumption of the model used throughout this book.

A safety culture may be defined as the attitude of employees within an organisation towards managing personal, corporate or environmental safety within their sphere of work.[4] While there are links between corporate, or organisational, culture and the way in which tasks are performed, including the development of safety consciousness within organisations,[5] national culture undoubtedly has a very strong influence upon both corporate culture and safety culture.[6] This being the case, the interpretation and implementation of the ISM Code will have as much to do with beliefs and values as it will with mandatory regulatory compliance.

By extension of the same reasoning, when assessing whether a specific company's SMS will be successful in achieving the stated objectives of the ISM Code, the person making the assessment on behalf of the flag state administration must make a subjective judgement based upon his own experience, education and training, all of which will inevitably be influenced by his or her own socio-cultural background.

The impact of culture upon the interpretation and implementation of safety regulations and development of a safety culture for two

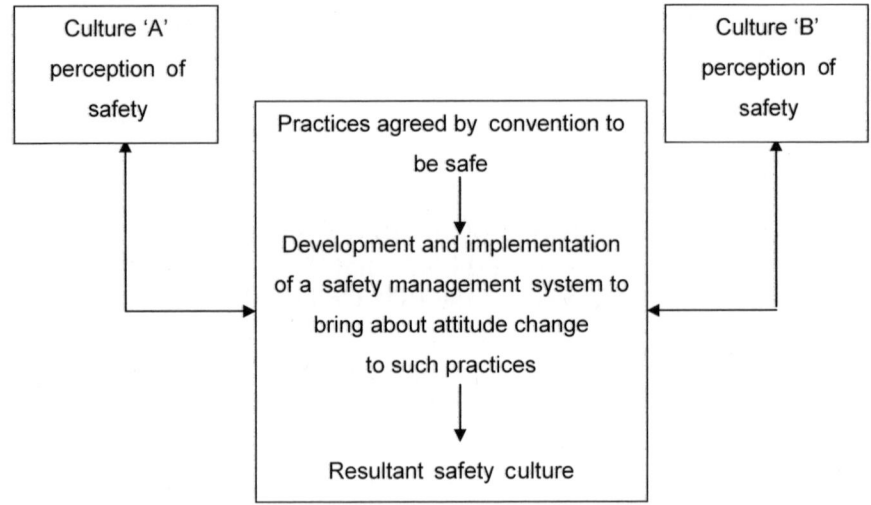

FIGURE 1. The impact of culture on safety models.

different cultures is illustrated in Figure 1, and the model may be extended to any number of cultures.

Shaw, on the other hand, is of the opinion that the notion of a safety culture *per se* is outdated and that safety should be viewed as an integral part of business, forming part of a company's overall risk management strategy. In the case of the ISM Code it is possible that there is no dichotomy in this respect, since the Code requires each ship-operating company to develop its own integrated SMS, i.e. a systematic approach to integrating safety into work planning and execution, encompassing protection of employees, the public, and the environment.

SAFETY CULTURE AND THE STCW CODE OBJECTIVES

While the objectives of the STCW Code are broadly in line with those of the ISM Code, the STCW Code does not use the term 'safety culture'. Instead, it endeavours to achieve its objectives by detailing those procedures which the parties to the Convention agreed were necessary to facilitate achievement of the Convention's objectives. No doubt, that in itself should in the long term be instrumental in helping to achieve a safety culture across the global shipping industry. However, since national culture undoubtedly has a very strong influence upon both corporate culture and safety culture,[6] how a particular nation state, shipping company or individual prioritises the procedures outlined in the STCW Code, what weight they give to specific requirements, will be influenced by their cultural backgrounds.

This being the case, then just as with the ISM Code, the interpretation and implementation of the STCW Code will have as much to do with beliefs and values as it will with mandatory regulatory compliance.

OBLIGATIONS ENGENDERED BY THE CODES

The establishment of the ISM and STCW Codes by means of international Conventions places obligations both legal and moral upon all states that are contracting parties to the Conventions to ensure that the provisions of the Codes are appropriately implemented within their jurisdictions.

Specific procedures within the STCW Code and specific provisions within the ISM Code relate to how the objectives of the Codes are to be achieved. The Codes therefore create a legal duty for ship-operating companies to address those specific procedures and those specific provisions, and for flag states to ensure that the obligation is fulfilled.

However, in addition to specific functional provisions such as the requirement to develop, implement and administer an SMS, the ISM Code also contains an objective declared in conceptual rather than specific terms: to provide an international standard for the safe management and operation of ships and for pollution prevention and thus ensure safety at sea, prevention of human injury or loss of life, and avoidance of damage to the marine environment. This is mirrored by the desire expressed in the STCW Code objectives to achieve and maintain the highest practicable standards for the safety of life and property at sea and in port and for the protection of the environment.

Although, as argued above, safety is a relative rather than an absolute concept and different positions may be held by different people on what constitutes 'safety', and although the objective of raising maritime safety standards is stated in conceptual terms, the stated objectives are fundamental to both Codes and create therefore

25

a moral obligation not only for ship-operating companies to satisfy the legal imperatives engendered by the Codes but also to administer their provisions in a manner that satisfies the spirit of the Codes and for flag states to ensure that this obligation is fulfilled.

SUMMARY

The chapter began by demonstrating that while the ISM and STCW Codes have similar objectives in so far as they both aim to improve maritime safety standards, there are fundamental differences in their approach, the ISM Code being prescriptive in outcome and stipulating only functional requirements while the STCW Code is prescriptive in process and stipulates precise procedural requirements.

The relationship between the two Codes on the one hand and both national culture and safety culture on the other hand was discussed, with emphasis on the impact of national culture upon the interpretation and implementation of safety regulations and subsequent development of a safety culture.

The moral and legal obligations engendered by the Codes were also identified, a topic that is touched upon again in Chapter 5 and investigated in more detail in Chapter 6.

3

Authority to Administer the ISM and STCW Codes

All I Ask is a Tall Ship and the Stars to Steer Her By

FLAG STATES AND INTERNATIONAL LAW

The nationality of ships

The idea of the secular nation state with sovereign rights was a philosophy initially developed by writers of the Renaissance period, prior to which the imperial concept prevailed.[1] The subsequent emergence of a plurality of nation states, each with its own sovereign authority, was paralleled by a need to regulate relationships between sovereign states and was the driving force behind the development of modern international law.

Today, the concept of the nation state, which sees all people in a geographical area as subject to one set of municipal laws, is the generally accepted political doctrine that governs international relations between states throughout the world.

With the development of the nation state as the sovereign power came the concept of nationality. However, the concept of the nationality of merchant ships was not fully developed until the end of the eighteenth century,[2] prior to which the principal point of legal reference was the nationality of a vessel's owner rather than the flag the vessel flew. Today, however, the fact that ships have nationality is recognised in international law under Article 90 of the 1982 United Nations Law of the Sea (LOS) Convention, which provides that every state whether coastal or land-locked has the right to sail ships under its flag on the high seas.

Article 91 of the 1982 LOS Convention further provides that:

- Each state shall fix the conditions for the grant of its nationality to ships, for the registration of ships in its territory, and for the right to fly its flag.
- Ships have the nationality of the state whose flag they are entitled to fly.
- There must exist a genuine link between the state and the ship.
- Each state shall issue to ships to which it has granted the right to fly its flag documents to that effect.

The 1982 LOS Convention came into force on 28 July 1996 and as of August 2006 had been ratified by 123 of the 151 UN member states participating in the agreement. By virtue of the provisions of the Convention the nationality of a ship is evidenced by the ship's papers and its flag, and is determined by registration with a nation state's ship registry, the conditions for which are laid down by the municipal law of the particular state and vary from state to state, with some states having less stringent conditions than others.

It has been suggested that the requirement for a genuine link between the state and the ship was introduced into the Convention in light of the judgement in the Nottebohm Case[4] in which the Court held *inter alia* that '... international law leaves it to each State to lay down the rules governing the grant of its own nationality ... On the other hand, a State cannot claim that the rules it has thus laid down are entitled to recognition by another State unless it has acted in conformity with this general aim of making the legal bond of nationality accord with the individual's genuine connection with the State which assumes the defence of its citizens by means of protection against other States.'

Thus, in accordance with Article 91 of the 1982 LOS Convention and the principle established in the Nottebohm case, the right to enjoy the privileges of the nationality of a state under international law arises only where there is a genuine link between the state and its citizen. But what constitutes a genuine link between a state and a ship is a matter of some debate within the shipping community.

In some states, simply incorporating a company in the particular state is sufficient to establish a genuine link as far as the municipal law of that state is concerned, although it is debatable whether that would be sufficient to satisfy public international law. However, the 1982 LOS Convention is silent on the consequences of there being no evidence of a genuine link between state and ship, and some maritime administrations have abandoned altogether any requirement for such a link as a condition for inclusion of vessels on their registers. The Marshall Islands Registry for example requires that vessels registered in the Republic of the Marshall Islands must be owned by a Marshall Islands citizen *or national of a qualified foreign maritime entity* (emphasis added). Foreign Maritime Entities are described in Section 3 of the Marshall Islands Vessel Registration and Mortgage Procedures (2005) as 'legal entities created under the laws of a jurisdiction other than the Marshall Islands that are eligible to own vessels when registered in the Marshall Islands pursuant to Section 119 of the BCA'.

Ship owners may be motivated to flag out their vessels to a foreign country where the link between the flag state and a ship is somewhat tenuous in order to reduce operating costs by taking advantage of that country's cheap registration fees, low or non-existent taxes and cheap labour costs. It is instructive to note that the majority of ships registered in the world's largest ship registries, Panama, Liberia, Honduras and the Marshall Islands, are owned by foreigners who register their ships in those states for fiscal, monetary and economic reasons, and even traditional maritime nations such as Norway now have second, or open registers, that have more favourable fiscal arrangements than their main registers.

When a ship registry offering such fiscal advantages also fails to demonstrate either the ability or the willingness to ensure that vessels on its register meet the operating, maintenance, and safety standards contained in international Conventions or to ensure that the operating companies employ crews that meet the basic standards of education and training required by international Conventions, then that state is said to be a flag of convenience country. As of August 2003, the International Transport Workers Federation[5] named 32 nation states as offering a flag of convenience.

The requirement for a genuine link was introduced partly in an attempt to limit the use of flags of convenience. That was considered necessary because it was apparent that some flag states were primarily interested in the revenue generated by ship registration and were not conscientious in fulfilling their obligations to ensure that ships flying their flags complied with the standards accepted by the flag state under the international Conventions to which it was party.

The revenue generated by a ship registry may be considerable. For example, between 1949 (when it was established) and 1999 the Liberian registry remitted around US$700 million to the Liberian government and in the year 2000 generated some US$18 million.[6] And it is interesting to note that the ship registries of a number of maritime administrations are not operated as part of the municipal administrative systems of their home states but are incorporated as companies that operate on a purely commercial basis and even their headquarters may be located in other states, a number being located in New York and London.[7] This of course does not necessarily mean that those administrations are any less conscientious than other maritime administrations in meeting their obligations created by international Conventions which their governments have ratified, but simply that commercial pressures do exist and that they may sometimes prevail over the niceties of public international law.

Clearly, if ship owners are free to select the state whose flag they wish to fly, and if individual nation states differ in the degree to which they ensure that ships flying their flags comply with internationally agreed regulations, then disparities may well arise between the standards to which different ships are maintained and operated, not only between ships sailing under different flags but also between ships flying the same flag but having different ownership.

Jurisdiction over vessels

Custom as a source of international law is recognised in Article 38(1) of the Statute of the International Court of Justice. However, just because a particular practice has become established over a period of time does not automatically mean that it has become a tenet of international law. In the Asylum Case[8] the court held that 'The Party which relies on a custom of this kind must prove that this custom is established in such a manner that it has become binding on the other Party.' This view was supported in the North Sea Continental Shelf Cases,[9] in which the court held that 'Not only must the acts concerned amount to a settled practice, but they must also be such, or carried out in such a way, as to be evidence of a belief that this practice is rendered as obligatory by the existence of a rule of law requiring it ... The States concerned must therefore feel that they are conforming to what amounts to a legal obligation. The frequency or even habitual character of the acts is not itself enough.'

With regard to jurisdiction over vessels, the fundamental principle under customary international law is that only the flag state may exercise jurisdiction over a vessel on the high seas and this principle is now embodied in Article 92(1) of the 1982 LOS Convention, which provides *inter alia* that:

> Ships shall sail under the flag of one State only and, save in exceptional circumstances provided for in international Treaties or in this Convention, shall be subject to its exclusive jurisdiction on the high seas.

The exceptional circumstances covered by the Convention refer to a limited number of criminal acts such as piracy, slave trading and unauthorised broadcasting.

But as discussed above, not all flag state administrations are equally effective or conscientious in ensuring that vessels flying their flags meet the minimum criteria established by international Conventions to which the states are contracting parties. And as each flag state is a sovereign power it enjoys sovereign immunity, there being no higher temporal authority to which it is answerable. As a UN agency, IMO can be regarded as a supranational body, but as pointed out by Lord Donaldson[10] the flag states are all voting members of IMO thus rendering improbable any possibility of censure of an errant flag state.

The question arises, therefore, as to whether states other than a vessel's flag state can enforce the obligations and duties accepted by the flag state as signatory to an internationally recognised IMO Convention.

PORT STATES AND INTERNATIONAL LAW

Under the 1982 LOS Convention, coastal states do have some rights over foreign flag vessels entering and transiting their territorial waters, particularly jurisdiction relating to navigation, defence and protection of the environment. These rights are provided for in Article 21 and Article 25 of the 1982 LOS Convention.

A port is generally considered to be part of a state's internal waters, and the general rule is that when a vessel voluntarily enters a foreign port it becomes subject to the coastal state's sovereignty since a state may enforce its national laws against foreign ships in its internal waters. This general rule was expressed in Wildenhus's

Case[11] by Wait CJ who held that 'It is part of the law of civilised nations that when a merchant vessel of one country enters the ports of another for the purposes of trade, it subjects itself to the law of the place to which it goes, unless by Treaty or otherwise the two countries have come to some different understanding or agreement.' By way of explanation he continued, 'As the owner has voluntarily taken of his vessels for his own private purposes to a place within the dominion of a Government other than his own, and from which he seeks protection during his stay, he owes that Government allegiance for the time being as is due for the protection to which he becomes entitled.'

However, the rule does not provide the coastal state with exclusive jurisdiction over the vessel. Matters that affect only the vessel and do not impact upon the port harbouring the vessel or the coastal state itself remain a matter for the flag state. Also, with regard to a ship's general condition and seaworthiness, the port state cannot force the vessel to meet standards higher than those recognised by the flag state. Provided the ship's papers and certificates are in order and the vessel presents no perceived environmental threat, the port state is not at liberty to carry out detailed inspections of the vessel. These matters were discussed in the Nimbus case[12] in the New Zealand Court of Appeal. William Rodman Sellers, master of the cutter *Nimbus* which was registered outside New Zealand, had refused to carry the radio and emergency beacon equipment required as a minimum by the New Zealand Director of Maritime Safety. Sellers subsequently sailed the vessel from Opua without obtaining the requisite port clearance. He was prosecuted for a breach of the New Zealand Maritime Transport Act 1994.

His appeal was dismissed in the High Court but he was granted leave to appeal to the Court of Appeal. Allowing the appeal, Keith J stated, 'Our conclusion on the relevant rules of international law is, accordingly, that a port state has no general power to unilaterally impose its own requirements on foreign ships relating to their construction, their safety and other equipment and their crewing if the requirements are to have effect on the high seas. Any requirements cannot go beyond those generally accepted, especially in the maritime Conventions and regulations ... In addition, any such port state powers relate only to those foreign ships which are in a hazardous state.'

The court also held that 'Legislation regulating maritime matters should be read in the context of the international law of the sea and, if possible, consistently with that law' and the United Nations Law of the Sea Convention 1982 was referenced several times. Under that Convention a vessel's flag state has primacy of jurisdiction over the vessel but Articles 211(3), 218 and 219 of the Convention provide a basis for port state jurisdiction, particularly with respect to pollution avoidance and containment, and to detain a vessel which is in violation of applicable international rules and standards relating to seaworthiness of vessels and thereby threatens to damage the marine environment (Article 219).

The first SOLAS Convention of 1914 also included a provision for a coastal state signatory to the Convention to inspect the papers of a ship flying the flag of a contracting party in order to ensure that they were valid and that the vessel therefore met the required standards. During recent years Port State Control provisions have been incorporated in a number of IMO Conventions. In particular Özçayir cites:

- SOLAS 74, reg. I/19, reg. IX/6 and reg. XI/4;
- Load Lines 66, Art.21;
- MARPOL 73 / 78, Arts. 5 and 6, reg. 8A of Annex I, reg.15 of Annex II, reg. 8 of Annex III and reg. 8 of Annex V;
- STCW 78, Art. X and reg. I/4;
- Tonnage 69, Art. 12.

Also, regulation 6 of Chapter IX of SOLAS 1974 refers specifically to Port State Control in respect of operational requirements with regard to the ISM Code and there are also provisions under the International Labour Organization No. 147, Merchant Shipping (Minimum Standards) Convention 1976 for port states to inspect vessels to ensure that minimum standards agreed to in the Convention are not being breached.

PORT STATE CONTROL

Primary responsibility for ensuring that a vessel complies with national and international regulations unquestionably lies with the ship's owner. It is the owner who is responsible for ensuring that the vessel is adequately maintained, properly crewed, sufficiently funded and well managed.

Secondary responsibility lies with the flag state. It is the duty of the flag state to ensure that the ship's owner is meeting the provisions of the state's municipal rules and the international Conventions to which the state is a contracting party. If the ship's owner is not meeting those provisions then the flag state has both the authority and the responsibility to insist that non-conformities are rectified, failing which it may remove the ship from its register, thus removing from the vessel the protection and privileges afforded by the state.

But as previously noted, not all flag state administrations are equally effective or conscientious, and since they are all voting members of IMO any possibility of censure of an errant flag state is rendered improbable.

A final defensive system against the operation of sub-standard shipping is provided by the littoral states, i.e. the coastal nations through whose territorial waters all vessels have the right under the 1982 LOS Convention to navigate unhindered. If a vessel founders in the territorial waters of a coastal state then it is the environment and citizens of the coastal state that are affected, not those of the flag State. This possibility has led to development of the Port State Control (PSC) system under which a percentage of ships entering a coastal state's ports are inspected. Any deficiencies found during the inspection are recorded and, if they are sufficiently serious, the vessel is detained until the deficiencies have been remedied.

To make the system as effective as possible, regional groupings of port states have combined to produce comprehensive regional policies. The first regional grouping was formed by 14 nation states that were signatories to the Paris Memorandum of Understanding (Paris MOU) 1982, to which a further five nations subsequently became signatories.

The Paris MOU, which is prescriptive in process and spells out in considerable detail precisely what a PSC inspector shall examine and what shall constitute a deficiency, served as a model for other coastal states interested in developing systems of Port State Control, with the result that the Tokyo MOU 1993 and the Caribbean MOU 1996 were subsequently established, both of which are recognised by the Paris MOU. A number of other regional groupings are currently developing MOUs for the Black Sea, the Indian Ocean, Latin America and the Mediterranean, while the USA and Australia each have their own unilateral Port State Control policies.

SUMMARY

This chapter looked at who is responsible for administering the ISM and STCW Codes. While that responsibility lies squarely upon the flag states, the responsibility for ensuring that vessels meet the minimum legal criteria for ensuring safety of life, property and the environment lies, in order of priority, with:

- The owner of the vessel.
- The manager/operator of the vessel.
- The vessel's flag state.
- Port states.

Because each of these entities has a different interest in the vessel and because each is subject to dissimilar cultural and socio-economic pressures, each entity may prioritise safety issues differently from the others. The cultural and socio-economic pressures acting upon each of these entities will be looked at in more detail in Chapter 4.

4

Safety Management

One Hand for the Ship and One for Yourself

SAFETY MANAGEMENT HIERARCHY

Le Guen[1] refers to a risk control hierarchy originally advocated by the Robens Committee[2] and since promoted by the Health and Safety Executive and the European Union. The concept of a hierarchy of norms associated with risk may be compared to Kelsen's pure theory of law[3] which argues that within a legal system there is a hierarchy of norms from the more abstract at the higher level to the more concrete at the practical level.

Kelsen's jurisprudential theory is relativistic in so far as it rejects the concept of there being only one single truth, holding instead that norms are relative to the individual or social group under consideration.[4] According to his theory a norm, by which Kelsen means the way things ought to be or the way a person ought to behave in particular circumstances, is valid only if it is authorised by another norm, which in turn must be authorised by a higher norm and so on until the basic norm is reached. In the case of a legal system the basic norm might be the written constitution of a state or, if the state has no written constitution, the norm-creating effect of custom.[5]

Similarly, it is possible to consider safety as a system or order of norms. By splitting up the concept of safety management in accordance with a number of hierarchal levels that, although comprehensively linked together, can be reviewed separately from each other, the constraints and pressures acting at each level can be identified and examined with respect to how they impact upon, or are influenced by, the development of safety management systems. For present purposes it is useful to categorise safety management under the following five headings, which provide a helpful analytic

36

framework for describing the emphasis adopted in the various safety management models described later on in this chapter and in developing a model of the working of the ISM Code in Chapter 8.

Level 1 – Imperative safety management

Imperative safety management refers to safety regulation at governmental level: negotiation of international treaties and conventions, and drafting of municipal laws relating to education, training, employment, protection of the environment and to safety and health of employees in the workplace (ES&H legislation).

Drafted at governmental level, either by national governments or by international governmental bodies such as agencies of the United Nations, safety regulation at this level may be considered a reflection of the will of the people and therefore constitutes the basic or constitutional norm of safety management: the norm from which all subsequent safety management stems

Level 2 – Institutional safety management

Institutional safety management refers to safety management at an industry level, how industry bodies interpret and organise implementation of the edicts of government, such as employment laws and safety and health (ES&H) legislation.

At this level the basic norm becomes a general norm and the fundamental idea of a duty of care starts to become concretised. Also at this level, the fundamental ideas of the basic norm become influenced by practicalities such as cultural norms and the economic circumstances of the companies, individuals or social groups that comprise the industry bodies under consideration.

Level 3 – Organisational safety management

Organisational safety management is the way in which individual companies respond to regulations emanating from the industry bodies implementing the edicts of government. This is safety management at the level of senior management.

At this level the general norm becomes a particular norm relating to a specific duty of care within the ambit of safety management and it is here that policies regarding safety management and risk management are developed and promulgated. Decisions are made regarding the crewing of company vessels, establishment of training programmes, the type and style of SMS that a company intends to follow and the resources that will be made available. Monitoring of the company's safety performance is carried out at this level utilising feedback received by senior management from operations personnel.

Level 4 – Operational safety

Operational safety is the practical organisation and implementation of safety. The particular norm becomes concretised. Staff are employed, training programmes are initiated, an SMS is developed, work procedures are formalised, and safety procedures are implemented ensuring that safe working practices are followed and suitable safety equipment is provided. Checks and balances are put in place to:

- identify hazards;
- minimise the possibility of undue risks being taken;
- provide feedback to senior management.

At this level the duty of care becomes a particular concretised norm in the form of a safety management system. Procedures have been promulgated and supervisors ensure that the procedures are followed.

Level 5 – Behavioural safety

Behavioural safety is safety at the level of the individual, the psychology of human behaviour in relation to operational safety in the workplace. This is the shipboard or factory floor level, the level at which work procedures are performed and at which safety procedures are aimed. At this level of the safety hierarchy the particular concretised norm has become a particular fully concretised norm.

By observing company policies, procedures and work instructions formalised in the company's SMS, the individual will be exercising on behalf of the company the duty of care at an operational level that the company legally or morally owes to people, property and the environment.

It is individuals who are responsible for ensuring that the duty of care is in fact exercised. But even though individuals are aware of their employers' policies and procedures they may sometimes choose to ignore some of those policies or procedures, just as some people choose to ignore particular laws. Why this should be so is a matter of much debate, but it may be that:

- the work instructions or safety procedures are ill defined; or
- the individuals are not sufficiently experienced, well-enough educated or adequately trained to carry out safely the tasks assigned to them; or
- there are over-riding economic considerations.

The first reason would be attributable to an organisational deficiency in the SMS caused at the organisational or operational level of safety management. The second reason would be an argument for more effective education and training. The third reason is looked at in some detail in Chapter 6.

Behavioural safety is particularly relevant in the area of education and training and is a key determinant in deciding whether stricter enforcement of existing regulations or greater emphasis on education and training is the better path to follow to ensure that the objectives of safety of life at sea, prevention of injury or loss of life and avoidance of damage to the environment are achieved throughout the shipping industry.

SAFETY MANAGEMENT MODELS

Central to the ISM Code is the development and implementation of a safety management system (SMS). It is expedient, therefore, to review the nature of safety management and the differing approaches that have been taken to its implementation as the concept has evolved.

Traditional approach

Early in the twentieth century attempts were made by The Safety First Movement to promote the concept of safety management by the introduction of safety committees with joint representation of management and workforce, and the appointment of safety officers to monitor adherence to the standards in the workplace.[6]

This approach, which operated at safety hierarchy levels 3, 4 and

5, was partially successful but did not promote proactive decision making, principally because the main purpose of the safety committees was to investigate accidents, the process and outcome of the investigation being steered by the preconceptions of the investigators about accident causation.[7] The investigation would attribute the cause of an accident to either:

- unsafe behaviour (unsafe act), in which case in order to prevent a recurrence of the unsafe act the committee would devise a rule forbidding such behaviour; or
- shortcomings in the working environment (unsafe conditions), in which case the committee would devise a technical solution to make the conditions safe or to protect people against the hazard.

The approach fitted well with the then prevalent theory of subjective and objective risk. The former related to the psychological dimension associated with a perceived danger while the latter related to the mathematical probability of the occurrence of an accident.

Watson[8] opposed the idea of there being a distinctive difference between subjective and objective risk, arguing that the theory was oversimplistic and valued abstract mathematical risk models above public opinion. Others[9] attacked the theory on the grounds that objective risk assessment cannot be free from an element of subjectivity, and in 1992 a Royal Society Report[10] concluded that subjective and objective risk assessment was no longer a mainstream theory. However, risk *per se* remains an important factor in safety management and is further investigated later in this chapter and in greater detail in Chapter 5.

Safety culture

Safety culture is a multi-faceted concept. For example, Sagen[11] terms the introduction of the ISM Code a paradigm shift in international ship operation and holds that the key factor in fulfilling the intentions and objectives of the ISM Code is the establishment of a safety culture in ship operation. Shaw,[12] however, defined safety culture as the attitude of employees within an organisation towards managing personal, corporate or environmental safety within their sphere of work.

The difference between Sagen's top-down perspective and Shaw's bottom-up standpoint may be due to the fact that the former appears to view safety culture as operating at level 2 of the safety hierarchy, while the latter sees it operating mainly at levels 3, 4 and 5 of the safety hierarchy, as with the traditional approach to safety management.

The ACSNI is of the opinion that safety culture is a sub-set of, or at least profoundly influenced by, the overall culture of an organisation, while Cooper[13] and Young[14] hold that far from reflecting shared values and beliefs, corporate culture is the result of conflict and alignment of many sub-cultures within an organisation, that it is a heterogeneous not a homogeneous phenomenon, leading Cooper to question whether industry-wide homogeneous safety cultures will ever arise, let alone a global one.

Undoubtedly, however, a nexus does exist between organisational culture and the way in which tasks are performed, including the development of safety consciousness within organisations.[15] But because national culture has an overwhelming influence upon both organisational culture and safety culture[16] it would be most unlikely for an industry-wide safety culture to be developed at level 2 of the safety management hierarchy, and even less likely for a global safety culture to be developed at level 1 of the hierarchy.

Safety cultures and safety management systems

Current management practice encourages the development of safety cultures within organisations and identifies safety management systems, a concept examined below, as an important element in developing safety cultures. The concept is illustrated in Figure 2 which shows the Reciprocal Safety Culture Model developed by Cooper.

Cooper's model, which adapts Bandura's[17] model of reciprocal determinism, comprises three interactive elements each of which impacts specifically upon one of the three component parts of a safety culture. The model also offers a means by which each element can be measured, and the various facets of the model can be elaborated upon and summed up as follows:

1. *Element – Person*
    Personnel selection – Person/job fit – Safety training – Competencies – Health assessments – Job satisfaction – Organisational commitment.
    Impacts upon – Safety climate

41

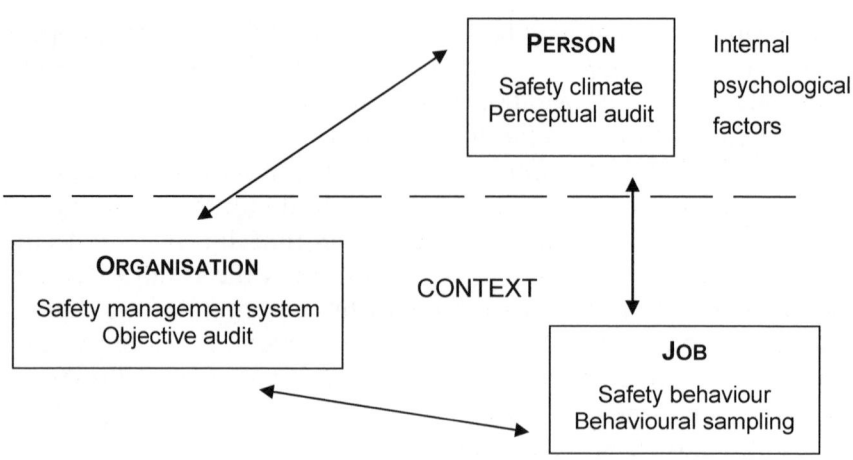

FIGURE 2. The Reciprocal Safety Culture Model.[13]

2. *Element – Organisation*
   Management commitment – Management actions – Communications – Allocation of resources – Emergency preparedness – Status of safety personnel – Policy and strategy development.
   Impacts upon – SMS

3. *Element – Job*
   Risk assessments – Required workplace – Standard operating procedures – Teamwork – Involvement in decision making – Person/Machine interface – Work environment – Work patterns.
   Impacts upon – Safety behaviour

Means of measurement

The means by which each of the three above elements can be measured are:

- Safety climate – Perceptual audit
- Safety management system – Objective audit
- Safety behaviour – Behavioural sampling

These means of measurement provide a useful framework for empirical research and will be utilised in the illustrative case studies that comprise the third part of this book.

INTEGRATED SAFETY MANAGEMENT SYSTEMS

Traditional safety law was prescriptive in process, specifying hazards and the preventive measures to be taken. A report by the Robens Committee – a safety hierarchy level 1 intervention that advocated self-regulation at safety levels 2 and 3 rather than resorting to legislation – recognised the shortcomings of the traditional approach to safety and recommended instead explicit policy objectives and effective organisation with clearly defined individual responsibilities. The ACSNI Human Factors Study Group Third Report expressed the opinion that what Robens called self-regulation would now be termed safety culture.

Interestingly, Robens' recommendation actually outlines the fundamental principles of a typical safety management system (SMS) that incorporates:

1. a policy that recognises management responsibility for safety, and
2. a safety organisation that:
   - identifies potential hazards and
   - apportions responsibility for dealing with those hazards.

Closed cycle safety management systems

Genn[18] has suggested that Robens' proposed self-regulation is successful in only a very limited number of companies, principally within firms that have clear, self-interested reasons for compliance, and that most employers will implement safety improvements only when detailed safety provisions are imposed upon them and the rules are enforced by an inspection system. Such an organisation may develop and implement an SMS such as that depicted in Figure 3 below and which may be termed a closed cycle model.

A closed cycle SMS is basic in design in so far as it is not incremental. Although the SMS may be fully integrated into the

43

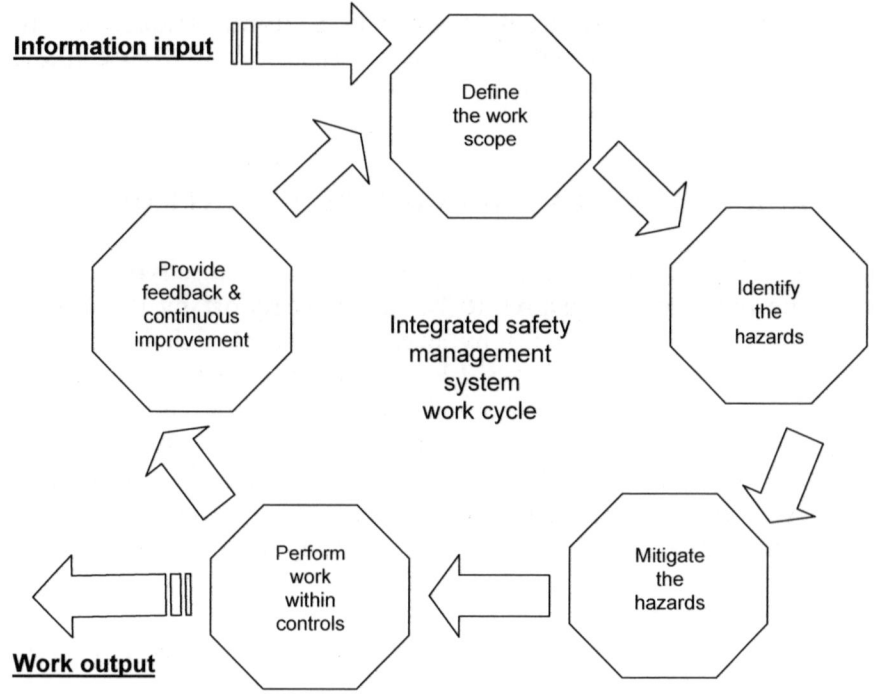

FIGURE 3. Model of a closed cycle safety management system.

organisation's work system, it is not open to outside influences and has only one point of feedback, suggesting that it has been introduced in order to achieve specific, limited goals. While feedback may enable such a system to mature it can grow in only two ways:

- by increasing the speed of the cycle; or
- by the addition of extra layers to the cycle, similar to the skins of an onion.

The first option is not realistic since the speed of the cycle will be constrained by the speed at which the work is performed. The second option is feasible but the system may become administratively cumbersome if too many layers are added.

An example of a closed cycle SMS may be found in Florida where state law (339.177 F.S.) requires the Department of Transportation to have an SMS in order to provide information needed to make informed decisions regarding the proper allocation of transportation resources. An SMS is broadly defined by the Florida State Safety Engineer's Office as the integration of the vehicle, the driver and the roadway elements into a comprehensive approach to solving highway safety problems.

The example given demonstrates the use of an integrated SMS to achieve a specific goal: to reduce the number and severity of traffic crashes by analysing feedback to ensure that all opportunities to improve road safety are:

- identified;
- considered, implemented when and where appropriate; and
- evaluated.

In terms of design, the system used by the Florida Department of Transport is a closed cycle SMS with single feedback. Although integrated into the organisation's work systems, the SMS operates on a closed cycle and is not intended to be significantly further developed.

Design of such systems is carried out by senior managers working at level 3 of the safety hierarchy, development and implementation of the systems are effected by middle management at level 4 of the hierarchy, and functional use of the systems is carried out by staff working at level 5 of the hierarchy. Feedback is between levels 5 and 4 of the safety hierarchy with no feedback at level 3 other than work output, which in the cited example is predominantly collection, interpretation and evaluation of data.

Incremental safety management systems

The phased repeal in the UK of outdated safety law – which was prescriptive in process – and the introduction by the UK Health and Safety Executive in 1992 of the Management of Health and Safety Regulations[19] – which are prescriptive in outcome, specifying the steps which an employer should take to identify hazards, assess risks, and select, implement and monitor preventive measures rather than specifying specific hazards and the preventive measures to be taken – have resulted in a more proactive approach to safety management, with safety management systems of an incremental nature integrated into an organisation's general operating framework.

The concept of an incremental integrated SMS with multiple lines of feedback is commensurate with a growing body of literature that recognises not only the personality factors and situational variables that affect both risk taking and risk-taking behaviour, but also the influence of the organisational structure itself.[20]

A model of an incremental SMS is illustrated in Figure 4 below. It is an open system with a comprehensive review mechanism at each level of operation, illustrated by the feedback loops in the figure, suggesting that the SMS is not only integrated into the organisation's work system but is also open to the influence of exogenous concerns such as public safety and environmental protection, and can be adapted accordingly.

In an incremental SMS, both management and operational tasks progress sequentially through five steps, each with its own feedback loop and each open to external influences, thus allowing the system to mature organically.

Employees and middle management working at levels 4 and 5 of the safety hierarchy carry out steps 1, 2, 3 and 4 of an integrated SMS while senior managers working at level 3 of the safety hierarchy

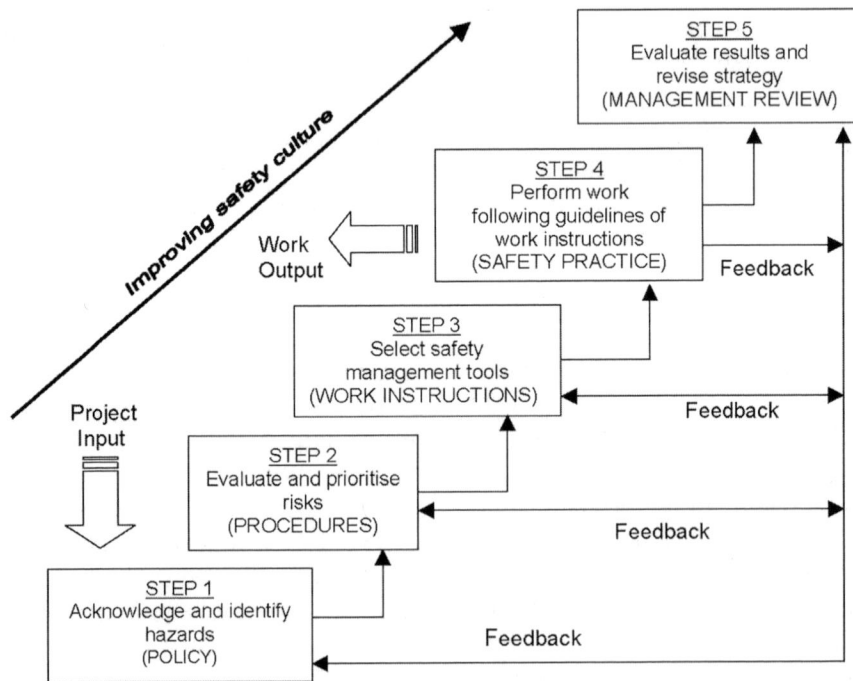

FIGURE 4. Model of an incremental safety management system.

not only influence the design of such systems but also actively engage in ensuring that the systems are effectively implemented and altered to accommodate changed circumstances as and when necessary, taking into consideration feedback received as indicated at step 5 in the diagram.

Comparison of SMS models

An important difference between the closed cycle and incremental SMS models is that the former involves single-loop learning while the latter embodies double-loop learning, a distinction first promulgated by Argyris[21] and later developed by Argyris with Schön.[22] According to this distinction, single-loop learning comprises detection and elimination of errors in accordance with given variables, while double-loop learning comprises changing the variables themselves.

Jankowicz[23] notes the distinction propounded by Argyris and Schön but highlights in addition that there are two different levels of language involved in the theory. The first apprehends control activities of an organisation at the operational level while the second is concerned with policy-making activities at the strategic level. These Jankowicz terms the subordinate and superordinate levels respectively, the former being associated with the learning organisation and the latter with the adaptive organisation.

Applying this concept to the two models of SMS illustrated in Figures 3 and 4 above, it may be seen that the closed cycle SMS has no identifiable superordinate level of control. The model is based completely upon controlling the development and implementation of safe working practices in an operational setting. Such an SMS will be essentially procedurally specific, comprising mainly company procedures and reporting formats. It involves single-loop learning.

In the incremental model, however, safety forms part of the organisation's risk management policy. While controls are in place at operational levels to monitor and develop safe working practices, feedback is also supplied from the subordinate level to the superordinate level at step 5, forming one of the inputs used by management to formulate strategic policy. Such an SMS involves double-loop learning and may be more outcome specific than a closed cycle model, incorporating not only procedural requirements but also policy guidelines.

An incremental SMS is most likely to be found in an organisation where management both empowers and encourages employees to exercise delegated authority and the SMS is seen as a way of systematically increasing the level of safety throughout the organisation.

The US Department of Energy's (DOE) Brookhaven National Laboratory (BNL), which conducts research in the physical, biomedical and environmental sciences, and energy technologies, provides an example of such an organisation. All DOE contracts incorporate an incremental SMS that BNL describes as combining all the elements of environment, safety and health into one ES&H system focused on accomplishing work safely rather than ES&H requirements and programmes for their own sake. BNL uses the acronym SIMPLE as an aid to remembering the five core functions of an incremental SMS:

- Define the Scope of work.
- Identify the hazards.
- Mitigate the hazards.
- Perform work within their controls.
- Lessons learned, feedback and continuous improvement.

According to BNL, the five core functions go hand in hand with the 'Eight Guiding Principles of Safety Management':

- Line management responsibility for safety.
- Clear roles and responsibilities.
- Competence commensurate with responsibilities.
- Balanced priorities.
- Identification of safety standards and requirements.
- Hazard controls tailored to the work being performed.
- Operations authorisation.
- Worker involvement.

There is a significant correlation between the BNL safety philosophy and that of DuPont Safety Resources[24] which is embodied in 11 principles, the first two of which are:

- All injuries are preventable.

- Management is responsible and accountable for preventing injuries.

The DuPont philosophy is currently fashionable in management circles, and large multinational companies such as Mobil Oil Corporation (now a part of the ExxonMobil Corporation) have had their operations audited by DuPont safety consultants. It is arguable, however, whether the first of DuPont's 11 guiding principles is attainable in practice. Theoretically it may be possible to completely prevent all injuries, but it is surely unrealistic to believe that it is possible in practice to avoid all accidents and their consequences. Indeed, one of the fundamental concepts of risk management is the tolerability of risk and that where it is impractical to completely eliminate a risk then controls are put in place to reduce the severity of the potential consequences of the risk should it eventuate.

MEASURING THE EFFECTIVENESS OF AN SMS

Whatever safety management system an organisation puts in place, whether it is prescriptive in process or prescriptive in outcome, closed cycle or incremental in nature, the objectives remain the same: ensuring the safety of personnel, property and the environment. Thus there is a need to be able to measure the effectiveness of safety systems.

The most common method of evaluating the effectiveness of an SMS is the recording of accidents, lost time incidents and hazardous occurrences. Fleming[25] endeavoured to go beyond this and develop a Safety Culture Maturity Model to assist organisations in the offshore oil and gas industry to:

- establish their current level of safety culture maturity; and
- identify the actions required to improve their culture.

In Fleming's draft model, which was not validated and therefore remains purely conceptual, organisations progress sequentially through the following five stages, moving from one stage to the next only when the strengths and weakness of each stage have been built upon or removed respectively:

Stage 1 – Emerging – Develop management commitment.

Stage 2 – Managing – Realise the importance of frontline staff and develop personal responsibility.

Stage 3 – Involving – Engage all staff to develop cooperation and commitment to improving safety.

Stage 4 – Cooperating – Develop consistency and fight complacency.

Stage 5 – Continually improving – Increasing consistency.

Measuring developments at each stage of Fleming's model would involve a large degree of qualitative and hence subjective judgement and it is difficult to imagine the concept of safety maturity without asking who decides the criteria that determine maturity. The model is therefore potentially very value laden. To mitigate this disadvantage Fleming proposed that the Safety Culture Maturity Model would be of relevance only to organisations that fulfil a number of criteria, including:

- An adequate SMS.
- Technical failures are not causing the majority of accidents.
- The company is compliant with health and safety law.
- Safety is not driven by the avoidance of prosecution but by the desire to prevent accidents.

However, if the term 'adequate' in the first point is meant to imply that the SMS is effective then this proposal overlooks two points:

- An SMS would be adequate only if it addressed the other three criteria.
- An effective SMS is self-improving and not a static system. It ensures that every hazardous occurrence and accident, whether or not it involves a lost time incident, is analysed to see what can be learned from the incident, what precautions can be put in place to prevent it happening again or to mitigate its impact if it cannot be prevented.

Although determination of the status of an organisation's safety culture may not be a legal requirement, it is something that a well-run company might well consider. Determination would involve the measurement of both qualitative and quantitative data and assessment would therefore best be achieved using a combination of perceptual audit, objective audit and behavioural sampling as proposed by Cooper. The possibility of measuring the maturity of an organisation's safety culture will be returned to in Part II of this book.

DEVELOPMENT AND IMPLEMETATION OF AN SMS

Development

When developing an integrated SMS for use in a shipping company the primary source of information concerning the structure of the system and the eventualities that it should encompass is the ISM Code and its Guidelines. The functional requirements for an SMS as envisaged by the ISM Code are laid down in Clause 1.4 of the Code as follows:

1. A safety and environmental protection policy;
2. Instructions and procedures to ensure safe operation of ships and protection of the environment in compliance with relevant international and flag State legislation;
3. Defined levels of authority and lines of communication between, and amongst, shore and shipboard personnel;
4. Procedures for reporting accidents and non-conformities with the provisions of this Code;
5. Procedures to prepare for and respond to emergency situations; and
6. Procedures for internal audits and management reviews.

Other sources of information would include, but not be limited to:

- Rules and regulations promulgated by flag state administrations.
- Rules and regulations promulgated by port state administrations.
- Rules and regulations promulgated by industry bodies.
- The company's own quality assurance system, comprising:
  - company policies;
  - operating procedures and work instructions;
  - feedback from management reviews.

Since Clause 1.2.3 of the ISM Code provides that the safety management objectives of the company should continuously improve safety management skills of personnel ashore and aboard ships, it is evident that the type of SMS envisaged by IMO should provide a systematic approach to integrating safety into work planning and execution, encompassing protection of employees, the public and the environment.

A feature of an incremental SMS is that it strives to continually improve the overall safety performance of the organisation in which it is embedded. However, that does not necessarily mean that only an incremental SMS will meet the IMO requirements since ongoing education and training within the parameters of either an incremental or a closed cycle SMS could also conceivably result in continuous improvement of the safety management skills of personnel.

So, in deciding what type of SMS is best suited to a particular shipping company various other factors also need to be taken into consideration, such as the size and type of vessels being operated, the number of crew members on board and the cultural and educational norms of the vessels' crews and shore-based management.

Implementation

Implementation of an integrated SMS in a shipping organisation involves validation of the SMS by an external industry body or government agency representing the flag state, which constitutes a level 2 intervention at level 3 of the safety management hierarchy.

Implementation also involves:

- Liaison between the shipping company and industry bodies such as flag state administrations and classification societies. This represents organisational safety at levels 3 and 4 of the safety hierarchy.
- Liaison between shore management and shipboard management. In terms of the safety hierarchy this represents an interface at levels 4 and 5 of the safety hierarchy: an interface between operational safety management and behavioural safety.
- A working relationship between the ship's senior officers and other members of the ship's complement. That is

behavioural safety, a particular fully concretised norm at level 5 of the safety hierarchy.

These interfaces between different levels of safety management frequently involve a cultural interface. The shore management and shipboard staff are, more often than not, of different nationalities[26] and even where they are of the same nationality then different organisational cultures may well prevail as a result of the different lifestyles and operating priorities that exist between shore management and shipboard staff. On board ship there is often a broad mix of nationalities, recent personal experience identifying nationals of five different nation states amongst a ship's staff comprising 12 persons.

However, while the input of personnel at levels 4 and 5 of the safety management hierarchy is directly relevant to the implementation of an SMS, its impact upon the design of an SMS is indirect, mainly by way of feedback to Level 3. From a management perspective implementation of an SMS is an organisational development problem rather than a qualitative one. A company determined to introduce a particular style of safety management on board its vessels has every means at its disposal to ensure that such a system is in fact implemented. There may be some resistance from ships' staff but a determined ship owner or operator can overcome that by using well-established techniques for dealing with resistance to change, although the practice may be more difficult than theory might suggest.

Safety management systems are designed and embedded in the policies and procedures of ship-operating companies at levels 3 and 4 of the safety hierarchy and it is at levels 4 and 5 that those policies and procedures are implemented.

CULTURALLY INFLUENCED VARIABLES

Holden[27] notes that culture can be used as an organising principle at different levels of human endeavour, citing as examples the international, the national, the regional, the organisational, the professional, the personal, each of which may be seen as a sphere of social-cultural interaction wherein culture is not merely determined by social norms but is influential in moulding social norms.

This concept maps conveniently onto the five-level safety hierarchy developed above. By acknowledging culture as a determining factor in the development and implementation of safety practices and then identifying the prevailing cultural dimensions (as per Hofstede[28]) at levels 3, 4 and 5 of the safety hierarchy, it is possible to establish whether they will have a positive (beneficial) effect upon the development and implementation of safety practices at those levels or whether their influence will be negative (counterproductive). That is, it would be possible to determine whether or not the prevailing cultural dimensions present particular risk factors to the establishment of safe working practices within an organisation. This concept will be examined in more detail in Chapter 5.

SUMMARY

This chapter began by identifying a five-level safety management hierarchy and went on to review two different approaches to safety management: the traditional concept, which was prescriptive in process, and the present-day notion of a safety culture, which tends to be prescriptive in outcome.

The concept of safety management systems was then explored, including factors influencing the effectiveness of an SMS and a comparison of closed cycle and incremental systems.

Quantitative and qualitative methods of measuring the effectiveness of an SMS were discussed and the possibility of measuring safety culture maturity was examined. Finally, the relationship between cultural dimensions and safety risk factors was touched upon. In the next chapter the relationship between culture and risk management will be examined more deeply.

5

Culture and Risk Management

Beliefs, Values and Heuristics

Safety, risk and the management of risk are closely associated and there are dimensions to both risk and the management of risk that are influenced by cultural and psychological factors.[1] This chapter explores those dimensions to establish how they might need to be accommodated in the cross-cultural safety management strategy of international, trans-national or global companies.

CULTURE

As noted in Chapter 1, Trompenaars defined culture as the means by which people communicate, perpetuate and develop their knowledge about attitudes towards life. Culture is the fabric of meaning in terms of which human beings interpret their experience and guide their action.

That is an excellent definition of culture in its broadest sense. However, culture needs to be more particularly defined when discussing specific social groupings, of which the following three are relevant to the area of study covered by this book.

- **National culture:** A set of deeply held, shared beliefs and values that underline the characteristics exhibited by groups of people within defined political boundaries.[2]
- **Corporate (organisational) culture:** Operational practices and attitudes of people within an organisation developed as a result of organisational edicts, custom and past practice. This definition synthesises definitions of organisational culture given by Pheysey[3] and Schein.[4]

- **Safety culture:** The attitude of employees within an organisation towards managing personal, corporate or environmental safety within their sphere of work.[5]

Other definitions of safety culture have been developed[6] but there is a general consensus that safety culture constitutes a proactive stance towards safety.[7]

CROSS-CULTURAL MANAGEMENT

There are links between corporate culture and the way tasks are performed, including the development of safety consciousness within organisations[8] and national culture is undoubtedly a compelling influence upon both corporate culture and safety culture.[9] This implies that in global, international or trans-national organisations effective safety management requires good cross-cultural management.

Holden[10] contends that cross-cultural management practitioners should not approach cultural diversity as a challenge to be dealt with but as a contextual factor in the administration of trans-national corporations, whilst Carnall (see note 8) sees national culture as a broader part of our affairs that requires an effective manager to display empathy, sensitivity to cultural differences and to be able to communicate in an intelligible fashion in a multi-cultural organisation and in cross-cultural situations.

Effective cross-cultural management therefore requires a manager to become familiar with the characteristics of the cultures of the people who may be regarded as stakeholders, both within and outside the organisation being managed.

MEASURING CULTURAL DIFFERENCES

Trompenaars[11] argues that it is possible to distinguish cultures from each other by observing the differences in shared meanings they expect and attribute to their environment. This accords with the findings of Hofstede who explored five dimensions along which national cultures differ from one another. These differences he termed:

- power-distance: a measure of the inequality in society;
- uncertainty avoidance: a measure of a society's tolerance of ambiguity;
- gender roles: a measure of the assertiveness of individuals in society;
- collectivism: a measure of the degree of individualism in society;
- long-term orientation: a measure of the time frame against which a society measures its values.

Hofstede illustrated how differences in these five dimensions could impact upon national affiliates of international organisations and it is therefore instructive to expand somewhat upon each of the dimensions.

Power-distance

Hofstede's Power-Distance Index (PDI), simply described, is a measure of the deference shown by people in a society to other people of different status within the same or a different society. It may also be seen as a measure of the inequality between people in authority and their inferiors, and the extent to which that is accepted.

Trompenaars is of the opinion that it is also a measure of how we accord status: whether it is ascribed by virtue of age, class, gender, education, etc. or attained as a result of an individual's personal achievements.

Uncertainty avoidance

Uncertainty Avoidance Index (UAI) is a measure of the degree to which an individual or society is comfortable with ambiguous situations, the extent to which they are willing to tolerate uncertainty.

Uncertainty avoidance should not be confused with risk aversion. The former refers to the avoidance of ambiguity in situations whereas the latter refers to an unwillingness to take a chance of achieving a possible outcome of an action with known outcome variables if there is a high degree of probability that the required outcome may not be achieved.

Gender roles

This dimension, which is also referred to as the masculinity versus femininity dimension and sometimes as the achievement versus relationship orientation, does not refer to the supremacy of males or females within a society but to the extent to which a culture stresses achievement or nurture.

Hofstede notes that different societies display to differing degrees characteristics that can be readily identified as male or female. A culture that displays primarily masculine traits emphasises ambition, wealth acquisition and differentiated gender roles whereas a culture that displays primarily feminine traits emphasises caring behaviour, sexual equality, environmental awareness and quality of life.

Collectivism/individualism

Collectivism as opposed to individualism is a measure of the degree to which people in a particular society believe that their duty is to act primarily for the greater good of that society rather than for their own personal advantage. It focuses on the relationship between the individual and the group within a society or an organisation.

Highly collectivist cultures believe the group is the most important unit and encourage primary loyalty to the group, decision-making based upon what is best for the group, and dependence upon organisations and institutions in the expectation that they will take care of the individual.

Highly individualistic cultures, on the other hand, see the individual as the most important unit and encourage people to be responsible for their own well-being and decision making based on individual needs.

Long-term orientation

To the four dimensions referred to above, originally identified by Inkeles and Levinson,[12] an additional dimension reflecting time orientation was subsequently added following studies in the Far East by Hofstede and Bond.[13] Originally termed Confucian dynamism by Bond, Hofstede's dimension uses the attitudes of different cultures to values such as perseverance, thrift and harmonious relationships, to compare and contrast long-term and short-term orientation in those cultures.

The significance of long-term orientation is that prior to its identification as an important dimension in an Oriental culture, studies had evaluated cultural dimensions from a purely Occidental standpoint, thus missing some of the important factors in cultures other than Western cultures.

PSYCHOLOGICAL DIMENSIONS

The fifth level of the safety management hierarchy described in Chapter 4 is behavioural safety, which is safety exercised at the level of the individual. Three psychological factors that have a strong bearing on behavioural safety are locus of control orientation, risk perception and cognitive biases, all of which may be affected by the educational, cultural and socio-economic background of an individual.

Locus of control

According to Heider[14] we observe the behaviour of others and then attribute causes to it. Heider's attribution theory holds that people see behaviour as being caused by either dispositional or situational factors, making a distinction between internally initiated actions and those resulting from reaction to external factors.

Other researchers[15] developed Heider's work, with Rotter concentrating on the situational and dispositional factors. Rotter identified in this an important personality trait which led him to develop a theory of locus of control, devising a personality test to measure an individual's locus of control orientation.

Locus of control orientation is a measure of a personal belief system about whether the outcome of one's actions is attributable to one's own actions and efforts (internal locus of control) or is contingent upon events outside one's personal control (external locus of control). This process can be represented diagrammatically, as shown in Figure 5.[16]

Subsequent studies using Rotter's inventory demonstrated that locus of control orientation is associated with an individual's socio-cultural background, family style and resources, and cultural stability.[17]

Trompenaars utilised Rotter's scale to measure the locus of

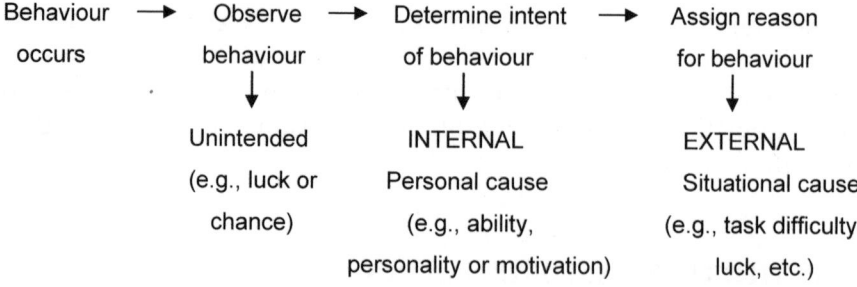

FIGURE 5. Diagrammatic representation of locus of control.[16]

control orientation of 30,000 managers in 48 different countries and found significant differences between geographical areas. Trompenaars concluded that national cultures are a very potent influence on individual loci of control to the extent that certain cultures have an overriding tendency towards internal control and others towards external control. He further contends, however, as does Mintzberg,[18] that for an organisation to succeed in business any significant corporate tendencies towards internal or external control need to be reconciled, a theme further developed by Mintzberg and Waters[19] in their model of deliberate and emergent strategy.

Other researchers[20] also found evidence to suggest links between culture and locus of control, while Bayne[21] established a significant positive correlation between individualism and internal locus of control although he found no evidence that collectivism was correlated to external locus of control.

Some researchers, notably Furnham,[22] disputed the findings of those researchers who argued that there is positive evidence to link culture and locus of control, citing inaccuracies in their studies and methodologies. However, on balance it would appear from the literature that the general consensus of opinion is that three cultural dimensions, power-distance, uncertainty avoidance and individualism, are indeed closely associated with locus of control such that the lower the PDI and UAI and the greater the degree of individualism of a person the stronger will be the tendency towards an internal locus of control orientation and *vice versa*.

A measure of an individual's locus of control therefore may be a rich source of information when studying behavioural safety at level 5 of the safety hierarchy. By measuring the locus of control of individuals and comparing it with their educational attainments, cultural backgrounds and work roles it might also be possible to discern whether or not locus of control has situational and chronological aspects, moving from external to internal as people gain knowledge and experience or gain promotion, i.e. as they gain greater control over their lives.

Such correlation would have a significant bearing upon deciding whether or not a greater emphasis on education and training would be successful in helping to establish common standards of safety throughout the shipping industry.

Risk perception

Research by Haley and Stumpf[23] provided some evidence that personality traits may predispose a manager to particular idiosyncratic biases, one of which is risk perception. Risk perception is a measure of the extent to which a person believes a particular course of action, or inaction, may result in harm occurring.

Faced with a situation involving risk, an individual makes a subjective assessment of probability based on data of limited validity, which are processed in accordance with heuristic principles.[24] Although the precise relationship between risk perception and risk behaviour is unclear[25], research by Simon (see note 17) identified three cognitive biases as lowering risk perception:

1. Overconfidence in the extent of one's knowledge.
2. Overconfidence in one's skills.
3. Belief in the law of small numbers.

The nature of these cognitive biases tends to indicate that they are the consequence of one or more of the following factors:

1. Lack of suitable education to be able to properly assess the risk.
2. Lack of adequate training to be able to properly assess the risk.
3. Lack of sufficient experience to be able to properly assess the risk.

This suggests that improved education and training should result in raising levels of risk perception, although it is unclear whether that would result in either improved safety attitudes or a reduction in accidents.

Cognitive biases

Cognitive biases are mental errors caused by the simplified mental processing of information as a result of fundamental limitations in human mental processes that not only cause us to perceive what we expect to perceive but also influence what information we remember and retrieve.[26]

Unlike other forms of bias such as organisational bias or bias resulting from self-interest, cognitive biases do not result from an intellectual predisposition towards a particular decision or judgement but from subconscious mental procedures for processing information based upon generalisations, simplifications and rules of thumb, which may in turn be based upon false premises.

When making decisions, managers use their experience and knowledge to review the information available to them. However, if they reach their decisions using inductive reasoning rather than normative judgements then cognitive biases may lead them to adopt inappropriate mental frameworks to evaluate the information, thus affecting the accuracy of any decision, regardless of the extent to which risks have to be estimated.

Research by Kahneman *et al* (see note 24) established the existence of the following three categories of definitive cognitive biases:

- **Input biases:** These occur when decision makers are selective in the data upon which they rely, giving some classes of data more weight than others.
- **Output biases:** These occur when decision makers fail to evaluate data appropriately, make guesses in the absence of data or supplement insufficient data with questionable data.
- **Operational biases:** These occur when decision makers either draw conclusions from inappropriate samples or jump to conclusions in the absence of data.

RISK AND THE MANAGEMENT OF RISK

The concept of risk is frequently associated with safety management models and covers a wide range of issues including self-reported risk taking, perceptions of risk within the workplace and attitudes towards risk and safety.[27]

Defining risk

In everyday speech the terms hazard and risk tend to be used interchangeably. However, Le Guen[28] argues that, pedantically speaking, hazard refers to the intrinsic propensity of any thing to cause harm, while risk refers to:

- The degree of probability of adverse consequences occurring as a result of a particular action or inaction taken in response to a potential hazard.
- A measure of the harm that may be done if those adverse consequences do in fact occur.

When people say that they are prepared to 'take a risk' then they are intimating not only that they have weighed up the chances of the adverse consequences occurring and the degree of harm that will result if those consequences do occur, but also that they are prepared to incur a chance of those adverse consequences occurring in expectation of a probable benefit.

Studies on the perception of risk carried out by a number of researchers[29] demonstrate that the concept of risk is strongly influenced by personal preferences and the values of the society in which we live. Whether or not a person, or an organisation, is prepared to take a risk depends upon a number of factors such as ethical and social considerations, economic and technical pressures, ignorance or knowledge of the potential adverse consequences and the degree of probability of those consequences occurring.

Based upon those considerations, preferences and values, individuals and organisations differentiate between risks they consider to be unacceptable, those that they perceive to be tolerable and those that are both negligible and broadly acceptable. Le Guen illustrates these factors as shown in Figure 6 below.

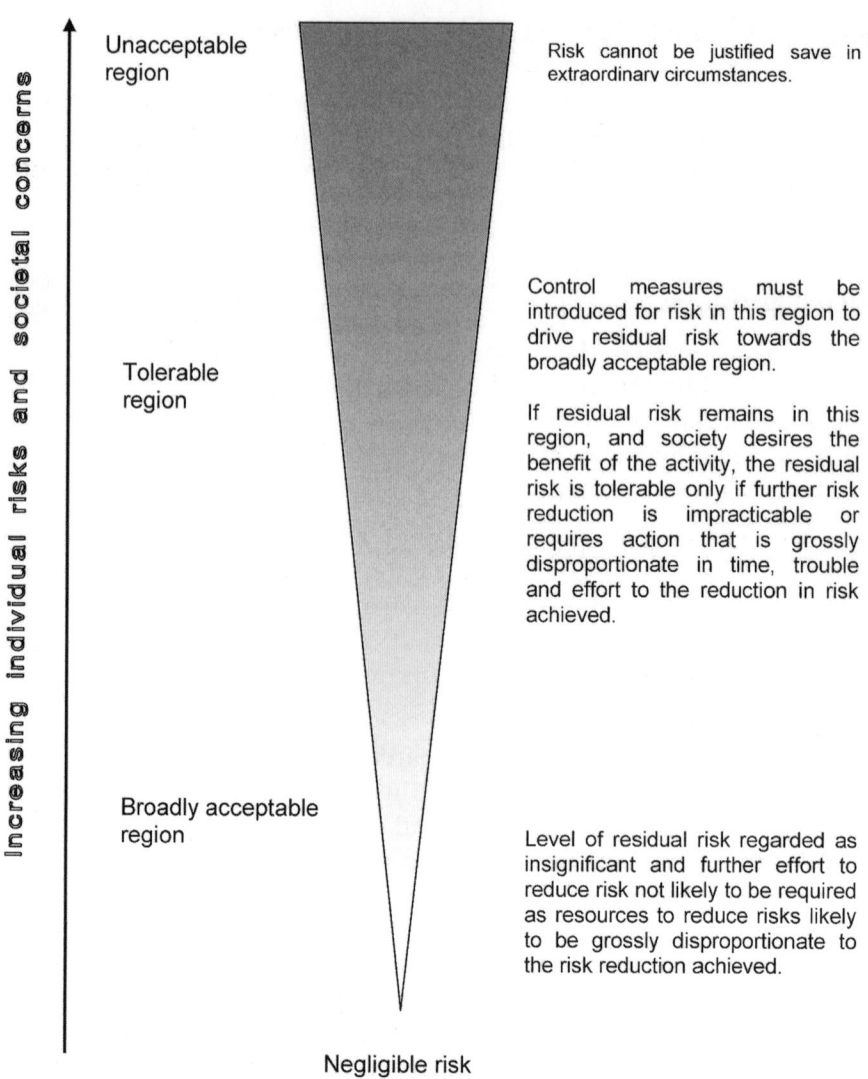

FIGURE 6. HSE criteria for the tolerability of risk.[28]

Managing risk

Risk management is a system of identifying potential hazards and their possible consequences then putting in place policies and procedures to deal with them. An effective risk management system enables a company to deal with strategic uncertainty by identifying threats and then either eliminating the associated risks or minimising the severity of the consequences should the risk eventuate.

Reflecting upon the diagram in Figure 6 it is clear that in order to arrive at an optimal risk management strategy a business must strike a balance between a cavalier attitude to risk on the one hand and obsessive risk aversion on the other.

As part of a study into entrepreneurship Hisrich *et al*[30] reviewed the concepts of risk taking and risk management. While noting that risk taking is indeed one aspect of being an entrepreneur, Hisrich concluded that despite many studies no conclusive causal relationships between entrepreneurship and a general propensity to take risks had been determined. Hisrich continued by describing specific risk reduction strategies in a context where risk refers to the probability and magnitude of downside financial loss. But the need to address hazards and reduce risk to an acceptable level by assessing the risks, evaluating the effectiveness of risk reduction measures and carrying out cost–benefit studies is applicable to all kinds of organisations not just entrepreneurial businesses, in which the ability to make measured risk estimates is well established. Indeed in many companies, particularly large multi-national corporations, the hazards and risks that need to be assessed go far beyond simple financial loss and involve more complex areas such as public perception and cultural acceptability, as discussed later in Chapter 7, which deals in some detail with culture, values, decision making and managing cultural diversity in the context of a global industry.

Embedded risk management systems

Shaw is of the opinion that safety should be viewed as an integral part of business, forming part of a company's overall risk management strategy. By extension of Shaw's reasoning, a company's corporate risk management system should form an embedded part of the company's SMS in order to increase its effectiveness. This should be a relatively straightforward process since the four fundamental steps in risk management[31] are in many respects similar to, and map well upon, the five steps of a safety management system (see Figures 3 and 4 above). The four fundamental steps in risk management are:

1. Acknowledge and identify risks.
2. Evaluate and prioritise risks.

3. Select risk management tools.
4. Evaluate results and revise strategies.

IDENTIFYING SALIENT ORGANISATIONAL RISK FACTORS

The risk factors with which this book is concerned are those associated with the development and implementation of an effective SMS in a shipping company rather than those associated with operational procedures *per se* since the latter are addressed by the operational procedures embedded within a company's SMS. The risk factors are therefore those associated with project development in a cross-cultural perspective for use by a company operating in a global environment.

As noted in Chapter 2 and further discussed below, at hierarchal safety management levels 3, 4 and 5 there exist three specific elements that are important in the development and implementation of an SMS and are the main determinants of its effectiveness:

- Leadership style;
- End user involvement; and
- Suitability of the SMS.

Therefore, in identifying the risks inherent in the development and implementation of an SMS in a shipping company, the risk factors should be reviewed in relation to those three salient elements, rejecting any items not relevant to the project and including under each head those items that are relevant. Any other factors that might affect the effectiveness of an SMS but which do not fit within one of the three main categories should also be reviewed and categorised separately.

Table 2 summarises the four risk factors considered to be important in the development and implementation of an SMS: leadership style, end user involvement, suitability of the SMS and any other factor affecting the adequacy or competency of human resources.

TABLE 2. SMS effectiveness: Determinant risk factors

Risk factor	Brief description

1. Leadership style results in:
  a. lack of top management commitment
    to a project
  b. conflict between user departments
	a. Lack of executive oversight, visible support and public endorsement. No active policy intervention.
	b. Serious differences in project goals, deliverables and design shared by different departments in an organisation.

2. Lack of adequate user involvement/ commitment	Active participation by end users is missing due to failure of developers to involve them. There is no commitment to deliverables and responsibilities.

3. Poorly drafted SMS resulting in misunderstanding of the objectives:
  a. by shore-based staff
  b. by sea-going staff
	a. Lack of executive and middle management understanding of ISM Code requirements and/or the cultural traits of the sea-going staff and operations/maintenance personnel.
	b. Consequent lack of understanding by sea-going staff of what is required of them.

4. Insufficient/inappropriate staffing or lack of required knowledge/skills in either shore-based or sea-going personnel	Risk that high expectations are mismatched with deliverables due to insufficient or inappropriate staffing, implying an inability to allocate a skilled workforce with suitable knowledge and/or experience in ship operations and maintenance, regardless of availability.

While all four risk factors are generally applicable to both the development and the implementation of an SMS, the first two are particularly relevant to its development and the second two are particularly relevant to its implementation.

Leadership style

Leadership style gives form and direction to organisational culture and it is commonly accepted throughout safety management literature that without senior management commitment there will be only token commitment to safety within any organisation.[32]

The introduction of an SMS, whether initially or to replace another SMS, involves an organisational change and subsequent monitoring of the system. To ensure that the new system is implemented effectively it is essential for senior management to consistently disconfirm patterns of behaviour not in keeping with the new organisational philosophy and consistently support any evidence of movement in the direction of the new assumptions.

End user involvement

Similarly, unless end users of a safety system are given a sense of ownership through participation and involvement in the development and operation of the system, then safety procedures may not be followed with any degree of enthusiasm, if at all, and authoritarian measures may need to be employed in order to achieve corporate safety goals.[33]

Suitability of the SMS

The type of SMS envisaged by IMO should provide a systematic approach to integrating safety into work planning and execution, encompassing protection of employees, the public and the environment.

The suitability of an organisation's SMS for achieving these objectives was discussed in some detail in Chapter 4, and from the discussion, particularly in the section entitled 'Development and implementation of an SMS', it is evident that if the style and type of SMS developed is incompatible with the type of vessels being operated, the number of crew members on board, the behavioural norms of the vessels' crews and the socio-cultural backgrounds of the shore management, then the suitability of the SMS will be questionable, leading to the possibility of misunderstandings and unsafe work practices.

Effectiveness of the SMS

With the first three risk factors suitably addressed there is still the possibility that other factors such as insufficient or inappropriate staffing may militate against the effectiveness of an organisation's SMS. Therefore, the fourth risk item comprises any other discernable factor affecting the adequacy and competency of human resources for the SMS to be effective and leading to the possibility of failure to achieve expectations.

THE IMPACT OF CULTURE ON THE RISK FACTORS

This book recognises safety management systems as subsystems of organisations, and that the observations of Hofstede on cultural differences and how they relate within organisations may well be applicable to safety management systems. Hence the integrated SMS, particularly the incremental model, is regarded as a highly value-laden system based upon subjective judgements that might well vary from one culture to another.

Therefore, having identified the risk factors associated with the development and implementation of an effective SMS, it is instructive to analyse how cultural dimensions might impact upon each risk factor, whether positively or negatively, i.e. whether the impact of a particular cultural dimension increases or decreases the likelihood of a particular risk eventuating.

For example, in situations where there is a high power-distance index, particularly where it is combined with a high degree of collectivism, then we may expect to see an organisation that is highly stratified and places great emphasis on the workforce being seen to obey the dictates of management.

Table 3, developed by McGill University,[34] presents a synopsis of the impact of Hofstede's cultural dimensions on management issues and the subsequent examples expand upon the synopsis given in the table, indicating how the impact of the cultural dimensions together with organisational weaknesses can increase or decrease the particular risk factors identified as being important in the development and implementation of an effective SMS.

TABLE 3. Impact of Hofstede's cultural dimensions on management issues (McGill University)[34]

Power distance	Small	Large
Organisational structure	Relatively flat	Hierarchical pyramid
Role of manager	Facilitator	Expert
Participative management	Possible	Not possible
Status symbols	Relatively unimportant	Very important
Importance of saving 'face'	Less important	Important
Uncertainty avoidance	**Weak**	**Strong**
Corporate plans	Seen as guidelines	Seen as important rules
Budgeting systems	Flexible	Inflexible
Control systems	Loose	Tight
Risk	Take	Avoid
Competition	Seen as advantageous	Seen as damaging
Individualism	**Collectivist**	**Individualist**
Decision making	Group consensus	Individuals look after selves
Organisational concern	Look after employees	Employees look after themselves
Reward systems	Group based	Individual/merit based
Ethics/Values	Particularism	Universalism
Masculinity/Femininity	**Feminine**	**Masculine**
Valued rewards	Quality of life	Money, performance
Networking	Important for relationships	Important for performance
Interpersonal focus	Maintaining relationships	Getting the task done
Basis for motivation	Service to others	Ambition – getting ahead

The following examples are not intended to be an exhaustive examination using all of the Hofstede/Trompenaars cultural dimensions in relation to all risk situations, but to provide illustrations of how the application of specific cultural dimensions to the four specific risk factors can be used to identify potential risks to the effectiveness of a shipping company's SMS and hence assist in enabling steps to be taken to negate or decrease the potential impact of the risks.

Risk factor 1

Leadership style results in:
a. lack of top management commitment to the project; and
b. conflict between user departments.

Two cultural dimensions identified by Hofstede considered to be very influential upon leadership style are power-distance and collectivism. In a society with a high power-distance culture there tends to be a rigid hierarchy, and societies with strong collectivism tend to be group orientated with members of one class tending to segregate themselves from those of another class. The end result of a combination of both dimensions is that subordinates do not question their superiors, and managers do not get involved in matters they have delegated to their staff.

While this satisfies societal goals of avoiding confrontation and maintaining an ostensible harmony in the workplace, it also results in top managers distancing themselves from middle and junior managers and therefore not being fully aware of the needs or progress of the project, both of which are essential for commitment.

Collectivism on its own can have either a positive or a negative effect on leadership style. In the context of a low power-distance culture it might have a positive influence because collectivism engenders, or at least emphasises, harmonious relationships and deters confrontation. But in a high power-distance context it might have a negative influence because groups of people within an organisation in such a culture will tend to favour their own group over another, and if this leads to conflict then neither group will defer to their senior manager as he or she will already have distanced himself/herself from the conflict.

Risk factor 2

Lack of adequate user involvement/commitment.

In a high power-distance culture there is a high degree of stratification and each layer of management distances itself from the layer below. This leads to a system of management by command rather than management by consultation. This does not lend itself to feedback analysis nor to a climate of user involvement or empowerment.

A high power-distance index therefore tends to increase the risk of lack of user involvement or commitment, whereas a lower power-distance index tends to have a positive effect and decrease the risk.

Risk factor 3

Poorly drafted SMS resulting in misunderstanding of the objectives:
a. by shore-based staff;
b. by sea-going staff.

In a society where a prevailing cultural trait is high power-distance, a senior manager will give an instruction and expect his or her orders to be followed without question. When the order is to develop and implement an SMS, middle managers and supervisors charged with developing the system may similarly remain aloof from their subordinates, in this case the end users of the system, and pretend to a higher level of expertise than they actually have simply in order to 'keep face' and demonstrate their authority. But without end user involvement and participation, the outcome may be an incomprehensible or poorly drafted SMS.

Power-distance therefore plays a significant role in the development of a well-drafted and effective SMS. A high power-distance index tends to increase the risk of lack of end user involvement and hence poor drafting of the SMS, whereas a lower power-distance index tends to have a positive effect and decrease the risk.

Risk factor 4

Failure to achieve expectations due to:
a. insufficient/inappropriate staffing; and
b. lack of required knowledge/skills in either shore-based or sea-going personnel.

Risk factor 4(a) relates to how well senior management understand the concept of a safety management system as required under the ISM Code and how committed they are to safety management. A truly committed senior management will ensure that adequate and properly trained staff are appointed to develop, implement and operationally monitor the company SMS. A senior management that only pays lip-service to safety management and does not truly understand what is required to develop, implement and sustain an effective SMS may result in insufficient and/or inappropriate staffing of the project.

Risk factor 4(b) is inherently connected with safety training, vocational and professional training and human resources management. It reflects the ability of a company to employ and develop a skilled workforce having suitable knowledge and experience in ship operations and maintenance, regardless of overall availability.

THE ROLE OF EDUCATION AND TRAINING

While newcomers to an organisation tend to learn patterns of behaviour from the people with whom they work,[35] not only at the level of skilled, semi-skilled and unskilled workers but also at professional and managerial levels,[36] there is a need for more formal training to ensure that existing personnel as well as new entrants to the shipping industry are both professionally competent and safety conscious.[37]

Companies that do not have an optimal risk management strategy are not operating as effectively as they could be and while most risk management strategies have to be cost effective, values are also important, because although risk management is frequently about containing or reducing the probability and magnitude of downside financial loss it also concerns human well-being, and it is here that education and training have a role.

As previously noted, in order to arrive at an optimal risk management strategy businesses must strike a balance between a cavalier attitude to risk on the one hand and obsessive risk aversion on the other. But a suitable balance cannot be achieved by safety training alone, for despite the notion that safety training will cure most ills in regard to accidents, evidence exists showing that it is not always effective.[38]

This may be related to the variability of the quality of training given or to the cultural attitudes of the trainees. It may also be due to a fundamental lack of competence by the trainee, since a person involved in deciding upon a specific course of action must have a certain level of knowledge in order to perceive the existence of a hazard and the associated degree of risk.[38]

Common standards of safety require common standards of competency, which in turn requires common standards of education and training. This does not mean safety training *per se* but ensuring that individuals have the technical knowledge to safely operate equipment under their control, such as ships and their machinery.[39]

Squire (see note 37) puts the point very succinctly when he says that the competence of a mariner will depend not only on good and effective education and training but also on aptitude, knowledge and understanding of the subject, on the availability of opportunities to develop skills and, ultimately, experience. Squire adds that competent people make the difference – they make the ship safe.

Concerns have been expressed about the adequacy of professional training of seafarers and its impact upon the levels of safety in shipping operations and practice.[40] However, Clause 6 of the ISM Code requires companies to ensure that Masters and seafarers on board their vessels are properly qualified and that training requirements are identified and provided. The Clause requires the company to:

- ensure that the Master is properly qualified to command (Cl. 6.1.1);
- ensure that each ship is manned with qualified and certificated seafarers (Cl. 6.2); and
- establish training requirements and ensure that it is provided (Cl. 6.5).

Shipping companies that interpret the Clause simply as a requirement to ensure that any seafarers they employ are properly qualified and in possession of the requisite certificates specified under the STCW 95 Code may have fulfilled their legal obligations. It is debatable, however, whether they have fulfilled their moral obligation to interpret the Clause within the spirit of the ISM Code as touched upon in Chapter 2 and further elaborated upon in Chapter 6. Regulations and standards are by definition minimum requirements. The spirit of the ISM Code is to introduce a genuine safety culture within the shipping industry and this implies training of staff beyond minimum stipulated requirements.

It is the maritime administrations of flag states that have both the responsibility and the authority for setting the required minimum standards of education and training for the award of their national certificates of competency. Historically, however, there has been wide divergence between flag states regarding their stipulated minimum standards of education and training. Recognition of this disparity prompted IMO to implement the 1995 revision of the 1978 STCW Convention, which entered fully into force in 2002 and contains provisions for an internationally agreed basic minimum standard of competency for the award of certificates of competency to seagoing personnel.

Lord Donaldson highlighted an important administrative difference between the ISM and STCW Conventions. Flag states signatory to the ISM Convention operate within a self-regulatory system under which the flag states certify to IMO that they are fully discharging their obligations under the Convention. The STCW Convention, on the other hand, contains a provision under which the signatory states to the Convention delegate to IMO the authority to assess whether or not a signatory is complying with its obligations under the Convention. How rigorously and with what ardour IMO monitors and enforces the provisions of the revised STCW Convention may well decide the effectiveness of the Convention in practice.

SUMMARY

This chapter explored:

- Risk and risk management and the way in which both are influenced by cultural and psychological factors.
- The link between national, organisational and safety cultures that establish the context in which cross-cultural management takes place.
- A number of cultural and psychological dimensions and how they could interact with four specific risk factors considered to be important in the development of an effective SMS.
- The role of education and training in relation to safety management.

The next chapter comprises a review of the legal, moral, cultural and economic pressures and constraints acting upon and within organisations implementing the ISM Code, and how those pressures and constraints might influence people employed in those organisations.

6

Pressures and Constraints Influencing ISM Code Implementation

Rights and Duties, Privileges and Obligations

OBLIGATIONS ENGENDERED BY THE ISM CODE

A Convention agreed between sovereign nations gains the force of municipal law when the sovereign nations introduce legislation in their individual parliaments adopting the Convention. When sufficient nations have adopted the Convention it becomes recognised as part of international law. But if laws are not obeyed, then instead of ensuring a basic minimum standard of performance within their area of concern, they become merely goals to be attained.

The ISM Code was incorporated within the international SOLAS Convention (1974) by the 1994 amendments to that Convention and as such is a part of international law. It is therefore important to examine what mechanisms are in place to ensure that the provisions of the Code are followed and what constraints and pressures might militate against observance of those provisions.

Safety is not an absolute but a variable that is influenced by the following factors which are determinants in the manner in which organisations and individuals respond to their obligations to observe the provisions of the ISM Code:

- The socio-economic context in which safety is to be observed.
- The hierarchal level at which safety is to be exercised.
- A legal imperative to act safely.
- A moral obligation to act safely.

What differentiates a legal duty from a moral obligation is a matter of jurisprudential philosophy, a detailed study of which is outside the scope of this book. However, for present purposes moral obligations may be understood to arise from the values and beliefs of a society, those factors which provide social cohesion, while legal duties are the result of rules of behaviour that are superimposed upon society and are supported by sanctions.

Harris[1] holds that most public figures believe that in addition to a legal duty to obey the law there is also a moral obligation to do so, identifying the following arguments for such a moral obligation:

1. Conceptually, if a regulation is recognised as 'law', it would be a contradiction to deny that it is binding or valid.
2. A duty exists because it is related to other moral concepts, specifically:
   a. gratitude;
   b. promise keeping;
   c. fairness;
   d. promoting the collective good.

Safety may also be regarded as a moral concept in so far as it concerns the relationship between a society's individuals, their property and their environment. While individual views regarding safety will be influenced by a person's own cultural and socio-economic background, there is a common thread running between all the different aspects of safety, and that is respect. Safety of the environment entails respect for the environment, safety of personnel entails respect for personnel, safety of property is a measure of respect for property and personal safety involves respect for oneself.[2]

Accidents happen because someone is careless, either in their acts or omissions, or in their evaluation of the risks posed by a particular hazard: they have paid insufficient attention to the consequences of their acts or their omissions. In other words, they have not acted with due diligence: they have not treated their environment, property under their control, their fellow workers or themselves with the respect they should have done.

It is a common tradition in Asia to take off one's shoes when entering a house, simply because it shows respect, both to the householder and to the house itself. Similarly, people entering a factory should ensure they are wearing the correct personal protective equipment, not only because it shows an awareness of the dangers of the factory environment but also because it demonstrates that the people entering the factory do not want to cause an accident involving either themselves or others: they are showing an awareness of their environment and respect for their own safety, the safety of others and the safety of the factory.

While the expression 'environmental safety' might be understood to refer either to the safety of our working environment or to the environment in general, in either case it requires us to show respect for our surroundings, whether they be our place of work, the countryside and oceans surrounding our place of work, the air into which our machines exhaust their fumes, the scenery around our homes or the beautiful scenery surrounding us when we go on vacation.

When engineers tune up a ship's engines after an overhaul they are not only complying with the engine manufacturer's maintenance procedures and helping to reduce fuel consumption, they are also helping to reduce exhaust emissions and hence atmospheric pollution, thereby demonstrating respect for the environment and consequently respect for other people who share that environment.

We have a duty to work and act safely, not simply because of legal and regulatory requirements but also because of a moral obligation to show respect for our co-workers, respect for property, respect for our communities and respect for ourselves. Respect is the basic societal norm of morality upon which the concept of safety is based, just as a duty of care is the basic norm upon which the legal enforcement of safety is based. Together they form the basic norm of the hierarchy of safety developed in Chapter 4.

Therefore, as was touched upon in Chapter 2 and again in Chapter 5, the existence of a moral obligation as well as a legal duty to exercise a duty of care obliges companies and flag states to observe the provisions of the ISM Code not only to the letter but also within the spirit of the Code, the former in recognition of a legal duty and the latter in recognition of a moral obligation.

Obligations at safety hierarchy level 1

MacLean[3] suggests that international law is relevant at three separate levels in international relations: the levels of cooperation, co-existence and conflict. With respect to the adoption and ratification of IMO Conventions the most appropriate level is that of cooperation and the obligations assumed by states that adopt the Conventions can be considered in the same light as obligations assumed under the law of contract.

At the very least, Treaties and Conventions agreed between nation states may be likened to relational contracts, the parties to which realise that it is not possible to provide for every possible contingency when the agreement is initially drafted and that later amendments or even complete redrafting may be necessary.

Unlike private contracts, agreements between nation states are not readily changed and thus full compliance by all parties may not be possible. Parties to an international agreement may accept therefore, tacitly or otherwise, that substantial compliance may be enough to satisfy the terms of the agreement.

Furmston[4] explains that obligations created by a contract are not all of equal importance and it is primarily for the parties to set their own value on the terms that they impose upon each other. He continues by pointing out that it is rare for contracting parties to express with any degree of precision what they have in their minds. Although Furmston is referring principally to the degree of emphasis each party might put upon particular contractual terms, the same argument can be applied to how contractual obligations are interpreted by parties having divergent cultures. Any differences of perception may be magnified when a degree of latitude is apparent in the need to fully comply with the terms of the contract.

Trompenaars[5] noted that in collectivist societies the relationship between contracting parties is more important than the detailed clauses of the contract itself and Ehrlich[6] acknowledged the dynamic nature of relational contracts when he distinguished formal sources of law from what he termed 'living law', by which he meant norms that prevail when the parties do not resort to litigation.

Similarly, Wacks[7] proposed that the law of contract is best understood through empirical studies rather than by studying formal sources of law, citing in support of his proposal a study by Macauley[8] which showed that the actual operation of commercial practice took precedence over the law of contract. A study by Beale and Dugdale[9] provided similar results, leading Wacks to conclude that the social context strongly influences the way in which contract law is practised.

Therefore, while the culture of the society in which a contract is developed will impact upon the nature and content of the contract, the culture of the society in which it is administered will influence the manner in which it is performed, particularly where the contract incorporates standards that can be determined by reference to community and commercial ethics that are heavily value laden, such as good faith, due care or fairness.

A society's law constitutes the chief bond between its culture and its organisation; it is the external manifestation of the embeddedness of the former in the latter.[10] Individuals follow the norms of the societies with which they are associated. They obey laws to the extent that their observance is in conformity with, and sanctioned by, the norms of the social group of which they are members.

There is also a political element to the introduction and implementation of regulations upon international commercial enterprises such as imposing the ISM Code upon the shipping industry. In the ordered societies of the developed world it may seem quite proper for nation states to agree among themselves a set of rules for the regulation of international shipping, but it gives rise to a deeply political question regarding the extent to which it is appropriate to impose minimum standards on people engaged in a commercial activity.

While nation states may have a moral obligation to encourage maritime safety and protection of the environment, commercial business is an economic affair the principal objective of which is to make profits and thereby increase the net present value of shareholders' wealth. Wealth creation is seen, particularly in developed capitalist societies, as not only an acceptable objective but also the preferred way of doing business and this raises the question of the right of governments to interfere with people's freedom to contract.

But whether or not governments do have any such right, the concept of freedom to contract has for many years had a limited application in business transactions in the developed world. In the United Kingdom for example, freedom to contract is restricted by numerous statutes regulating various aspects of commerce such as contract exemption clauses (*Unfair Contract Terms Act, 1977*) employment (*Employment Protection [Consolidation] Act 1978*) and the health and safety of employees in the workplace (*Health and Safety at Work etc Act 1974*).

The phenomenon of the law intervening to restrict and regulate the freedom to contract and provide an equitable balance between the contracting parties is not solely a product of post-industrial philosophy: it was clearly evident in Britain for example in the Middle Ages when market courts were established to administer the law merchant and deter unethical trading.[11]

Today, most societies in developed countries that support the concept of commercial profitability have laws governing socially unacceptable ways of maximising profit, such as for example by employing child labour, paying low salaries and condoning unsafe working practices. But that is not the case in many developing countries where business ethics do tolerate child labour, low salaries and unsafe working conditions.

Yet another approach to commerce and the role of the state was taken by the command societies of the Soviet Union, China and Eastern Europe, which looked upon personal wealth creation as morally repugnant and promoted instead the concept of collective wealth creation.[12] It is debatable whether the state as employer in centrally controlled economies was less concerned about the health and safety, employment rights and working conditions of the individual than was the state as legislator in capitalist economies.

But whether profit maximisation is seen as an individual or a collective notion, one factor militates in favour of imposing the rules contained in the ISM Code: the introduction of minimum standards throughout the entire shipping industry would be an equitable measure that would establish a basic norm, a 'level playing field', for companies involved in international shipping operations.

People generally will agree, or at least concede, that commercial dealings should be conducted fairly, either because they agree with the proposition or because they are not willing to express opposition. But Cooke,[13] in discussing the concept of fairness as a criterion for judicial decision making, contends that 'For fairness to work as an effective criterion it is necessary that the society [has] a more-or-less common set of values and that this value is high amongst them'. This presents an argument in favour of stronger policing of existing regulations by Port State Control inspectorates since the global diversity of cultures must infer a diversity of values.

From the foregoing discussion, the pressures and constraints to be found at level 1 of the safety hierarchy may be briefly summarised as:

- The way in which the political and legal institutions have been shaped by the culture and history of individual nations.

- The manner in which a society views law, whether as a body of regulations to be obeyed or as identifiable targets to be achieved.
- The degree to which nation states are industrially developed and are prepared to accept what developed nations consider to be fair and just standards.

Obligations at safety hierarchy level 2

When considering the ISM Code, the primary organisations at level 2 in the safety hierarchy are the flag state maritime administrations. They have both the responsibility and the authority to administer national and international maritime rules and regulations that have been sanctioned by their respective governments.

However, they are also under pressure to increase the amount of shipping flying their national flags, thereby increasing national revenue and, as discussed in Chapter 3, this has given rise to a number of ship registries paying only lip-service to the implementation of maritime rules and regulations, allowing unscrupulous ship owners to operate what has become known colloquially in the marine industry as sub-standard tonnage, i.e. vessels which neither comply with the requirements of international Conventions nor meet the requirements of the constituent members of the International Association of Classification Societies.

Introduction of the ISM Code and policing of ship standards by Port State Control bodies have gone some way towards counterbalancing this tendency. As noted in Chapters 2 and 3, under the provisions of international law flag state administrations are obliged to ensure that ships on their registers are operated in conformity with the ISM Code, and the practice of policing of ship standards by Port State Control bodies is one means of ensuring that the flag states do in fact meet their obligations.

The pressures and constraints to be found at level 2 of the safety hierarchy are occasioned principally by the same agencies as those at level 1 and may be briefly summarised as:

- A legal and moral obligation by maritime administrations to ensure that the provisions of the ISM Code are implemented and enforced.

- Economic obligation of maritime administrations to ensure that ship owners will continue to provide revenue by registering ships with them.
- Public pressure, particularly in countries where a large number of people are engaged in the shipping industry, to remain competitive and ensure that employment within the industry is not jeopardised by inflationary costs.
- Public pressure on maritime administrations to ensure that vessels entering their waters do not cause damage or pollution of the environment.

Obligations at safety hierarchy level 3

When a company registers a vessel in a particular state the act of registration signifies the company's agreement to observe the state's domestic laws including obligations arising from international Conventions which the state has adopted. This imposes a legal obligation on the company to run their operations in accordance with the provisions of such Conventions, for example the provisions of the ISM Code.

The judgement in the case of M/V *Eurasian Dream*[14] highlights the fact that shipping companies also have obligations to corporate stakeholders, i.e. those who are directly or indirectly affected by corporate decision making such as charterers, shareholders, cargo interests and insurance companies, to run the company and the vessels it operates with due diligence and in accordance with the latest industry standards.

M/V *Eurasian Dream* was a car carrier discharging its cargo in Sharjah when one of the vehicles caught fire. That led to the vessel being abandoned and destroyed by the fire. The judge held that the claimants, who were the cargo interests, had proved that the carrier had breached Article III of The Hague/Hague-Visby Rules because the vessel was unseaworthy due to 'numerous failures and errors of judgement that amounted to professional negligence in respect of the provision of equipment, competent master and crew and adequate documentation'.

Although at the time of the fire the vessel was not ISM certified nor was it required to be, the ship had been provided with copies of the managing company's ISM procedural documentation and was subject to the same company procedures as all other vessels in the fleet. However, the judge was highly critical of the SMS documentation on board because it was not ship-specific and hence largely inappropriate and irrelevant. He was also very critical of the company's training policies, describing the vessel's Master as a 'car carrier novice'.

The judge found that the ship management company had failed to exercise due diligence based upon reasonable standards and practices of the industry at the time and the fact that although the vessel had onboard much of the ship manager's ISM documentation their SMS was utterly deficient.

From the foregoing and the discussions in Chapter 2 and Chapter 6 above, it is apparent that the following constraints and pressures bear upon senior decision makers in shipping organisations:

- There is both a moral and a legal obligation to run their operations in accordance with the provisions of the ISM Code.
- There are economic factors to take into consideration which may be reflected in budgetary constraints with regard to safety, training and the quality of ship maintenance.
- The type and style of safety management system developed and implemented will reflect:
  a. the prevailing cultural norms of the decision makers;
  b. the organisation's corporate culture;
  c. management style;
  d. managerial competence.

Obligations at safety hierarchy level 4

The ISM Code is aimed principally at the shore-based management of shipping companies rather than the sea-going staff. It does however address the ship/shore interface at levels 4 and 5 of the safety hierarchy, in so far as the functional requirements for an SMS defined in Clause 1.4 of the Code require:

- Safety and environmental policies developed by shore-based management to be conveyed to sea-going staff by way of instructions and procedures.
- Defined levels of authority and lines of communication between, and among, shore and shipboard personnel.

The requirement for defined levels of authority and lines of communication is expanded upon in Clauses 4 and 5 of the Code, which:

- require the appointment of a designated person ashore
  (DPA) to provide a link between the company and those on
  board, and
- outline the Master's responsibility and authority for
  implementing, reviewing and reporting upon the shipboard
  provisions of the company's SMS.

Once a decision has been made by senior management to comply
with the provisions of the ISM Code, it is the responsibility of middle
management and supervisory staff to give effect to that decision,
utilising the resources available to them, following the guidelines of
the ISM Code and taking into consideration the prevailing organisational and cultural norms, both ashore and on board ship.

The constraints and pressures operating at safety hierarchy level 4
therefore bear principally upon middle management and supervisory
staff who, in developing and implementing an SMS, will be influenced not only by the ISM Code guidelines but also:

- the extent of available resources;
- prevailing organisational and cultural norms; and
- their own competencies and understanding and that of the
  sea-going staff.

Obligations at safety hierarchy level 5

The most important aspect of safety at level 5 of the safety hierarchy
is behavioural safety, safety at the level of the individual, the psychology of human behaviour in relation to the problems of safety in
the workplace. It is particularly relevant in the exercise of a duty of
care, because although the law may recognise that companies and
organisations owe a duty of care to others, it is individuals who have
de facto responsibility for ensuring that the duty is in fact exercised.

The Master of a merchant ship is both the legal representative and
agent of the ship owner. He or she is therefore in a position of trust
and under an obligation to act in the interests of the ship owner, but
always within the parameters of the law. The Master of a merchant

ship is also responsible for the safety of the ship, its cargo and its crew.

Occasions may arise when the interests of the owner conflict with the requirements of safety and in such circumstances the Master should act in the interests of safety and would be under pressure to do so, the obligation being both a moral one and a legal one.

In a well-found company, policies and safety procedures contained within the SMS would be sufficiently detailed to provide guidance in the event of a conflict arising between the interests of the ship's owner and the obligations of the ship's Master. In less conscientious companies however, if such a situation were to arise the Master of the vessel might well find himself under considerable pressure to put the interests of the owner before matters of safety.

In the former type of company there would be pressure on the vessel's Master and senior officers to ensure that the company's policies, procedures and SMS were fully implemented. In the latter type of company the policy and procedure manuals and the SMS documentation would all be found on board the vessel but there would be no pressure from the company for the Master to ensure the implementation of their contents. Rather he would be given little guidance and would be expected to use his initiative and act in the company's best interests, particularly its economic interests. This appeared to be the situation in the case of the *Eurasian Dream* outlined above. The Master was new not only to the vessel but also to the ship management company. He was not familiar with the vessel's fire-fighting systems, had apparently received no training relevant to car carriers and had been instructed simply to read the hundred or so manuals on board.

But even in a well-found and conscientious company there may be gaps in the company's operational procedures as was noted in the case of Davis v Stena Line.[15] The case concerned the circumstances surrounding the death by drowning of a passenger who went overboard from the ferry M/V *Koningin Beatrix* on the morning of 29 October 2000. The judge noted that 'In October 2000, Stena's current Standing Orders and Operational Procedures Manual contained no guidance and no specific operational procedure for rescuing a man overboard in the event that it was not possible to launch the vessel's own rescue boat.'

However, gaps in company procedures are not the only reason why correct procedures are sometimes not followed. Safety, like law, comprises a system of rules and regulations and although individuals may be aware of their employer's policies and procedures they may sometimes choose to ignore some of those policies or procedures, just as some people choose to ignore particular laws. Why this should be so is a matter of debate, not only among management psychologists and legal philosophers but also among people charged with ensuring safety directives are not only promulgated but also acted upon. On the one hand it may be due to individual personality traits or psychological dimensions: on the other hand it may reflect management pressure to ignore company safety rules in return for a perceived corporate benefit.

However, non-observance of safety rules and regulations at this operational level may be due to the person involved simply not being aware of the prevailing rules and regulations, which may in turn be due to the type and style of the company's SMS, lack of suitable training or sheer incompetence on the part of the individual.

Drawing upon the foregoing discussion, the most salient constraints and pressures acting at this level of the safety hierarchy arise from:

- obligations to follow company policies and procedures;
- educational and training norms as reflected in the competency of individual seafarers;
- psychological dimensions of individuals.

PENALTIES FOR NON-COMPLIANCE

Why people obey the law is a matter of much debate among writers of books about jurisprudence but is also relevant when considering how best to enforce the provisions of international Conventions such as the ISM Code.

In the field of jurisprudence Rawls' social contract theory of justice[16] argues that there is a moral obligation to obey the law *per se*, while advocates of the utilitarian theory argue that people obey laws to the extent that the law is in conformity with, and sanctioned by, the norms of the social groups of which they are members. Legal positivists such as Kelsen[17] contend that duty and sanction exist separately but if a valid law exists then it is binding and people obey the law because of the fear of sanctions that will be incurred if they do not obey the law. These jurisprudential arguments revolve around three possibilities:

- People obey the law because they have a moral duty to do so.
- People obey the law for fear of sanctions.
- People obey the law because of perceived benefits.

What is also of interest in the present study is how people conceive the strictures of the law: whether as strict rules to be obeyed or as goals to be achieved. This may well vary between people with different cultural backgrounds and is pertinent to the introduction by a shipping company of an integrated SMS in compliance with the ISM Code, since the way in which the provisions of the SMS are interpreted and to what extent they are followed may well determine whether or not a vessel passes or fails a safety audit by either a flag state or a port state inspectorate.

There are no stipulated penalties for non-observance of the provisions of the ISM Code, only the possibility that failure to observe its provisions may result in detention of a vessel, withdrawal of its safety management certificate (SMC) or, in the case of a serious breach, withdrawal of the operating company's document of compliance (DOC) thus rendering the company unfit to manage vessels. Clause 3.12.1 of the ISM Code states 'The Company is responsible for determining and initiating the corrective action needed to correct a non-conformity or to correct the cause of the non-conformity. Failure to correct non-conformities with specific requirements of the ISM Code may affect the validity of the Document of Compliance and the related Safety Management Certificate.'

Only the flag state that issued them can withdraw a vessel's SMC or an operating company's DOC. However, as discussed in Chapter 3, a port state may detain a vessel if it considers the condition of the vessel to be such that it poses a threat to the safety of the nation's environment or citizens, although Lord Donaldson[18] is of the opinion that detention of a vessel may not be overly burdensome for its owners, presenting more of an inconvenience than a deterrent. Donaldson argues that in order to provide an incentive for owners to comply with the provisions of the ISM Code and act as a deterrent against flouting its provisions, detention should be both costly for the errant ship owner and highly profitable for the port authority compelled to provide harbour space for the vessel until the noted deficiencies have been made good.

This argument, however, is somewhat contentious for two reasons. First, detention of a vessel becomes more than a mere inconvenience to owners if as a result of the detention the revenue-earning potential of the vessel is affected, as for example when a vessel misses its lay-days and consequently loses a charter. Second, allowing a port authority to impose high charges on a detained vessel pre-supposes that the ship owner is the guilty party and the port state control inspector is the virtuous party. Such a situation could lead to corrupt officials detaining vessels of innocent parties.

ECONOMIC CONSIDERATIONS OF NON-COMPLIANCE

There is a widely held presumption that ship owners can obtain competitive advantages by deliberately failing to observe applicable international rules and standards pertinent to the safe operation of ships. The belief relies upon an assumption that considerable scope exists for ship owners to deliberately avoid compliance with international rules and standards which govern safety and pollution prevention. However, this is difficult to reconcile with the ever-increasing number of surveys that ships are forced to undergo, such as:

- flag state inspections;
- port state inspections;
- classification society surveys;
- charterers' inspections;
- P&I club surveys;
- financiers' inspections.

In support of its contention that economic pressures have forced an increasing number of ship owners onto a survival footing, characterised by cost-saving initiatives and expenditure cutbacks on safety-related maintenance with the risk of violating international rules and standards, the OECD[19] cites a study by Intertanko showing a decrease in the amount of time spent in dry-dock by very large crude carriers (VLCCs). The report notes a reduction of 27.5% in dry-dockings between 1991 and 1994 with the average number of days spent in dry-dock falling from 24 to 22. However, the report fails to consider pertinent factors such as prevailing market conditions during that period, the five-year docking cycle of VLCCs, increases of efficiency in shipyards due to technological advances, the improved performance of underwater hull coatings enabling intermediate surveys to be carried out while vessels are afloat, and the possibility of using afloat repair squads to carry out essential repairs and maintenance during ballast passages.

But it would also be naive to believe that any ship owner would maintain their vessels to a very high standard unless they saw an economic advantage in so doing. Owners operate and maintain their vessels in compliance with regulatory requirements with regard to four overriding criteria:

- the earning potential of the vessel;
- charterers' requirements;
- the resale value of the vessel;
- market trends.

These four criteria dictate the economic viability of the vessel. It would not make economic sense for an owner to spend more than necessary to operate or maintain a vessel to meet charterers' requirements if it meant that expenditure exceeded the earning potential of the vessel. By the same token, it would not make economic sense for an owner to spend less than needed to operate and maintain a vessel to charterers' requirements if the under-expenditure compromised the earning potential of the vessel. As Shaw[20] points out, in order for organisations to operate efficiently, financially and otherwise, it is necessary for them to strike a balance between having on the one hand no safety standards at all and having on the other hand excessive and obsessive safety standards.

This is not to deny that some ship owners spend more on maintenance and operations than others but the reasons are far more complex than generally supposed. The major oil companies tend to maintain their vessels within the level identified by the OECD as that of good practice, i.e. a high level of expenditure adopted by a minority of ship owners. There is one over-riding reason for that: reputable ship owners, particularly major oil companies, know that accidents cost money and may create bad publicity, as was the case when the oil tanker M/T *Exxon Valdez* went aground on Bligh Reef,

Alaska, on 24 March 1989, spilling 38,800 metric tonnes of crude oil into Prince William Sound. They are therefore concerned to ensure that their vessels are maintained and operated to a high standard, thus reducing the risk of an accident occurring, subsequent actions for damages, and the negative impact of any ensuing publicity.

Some owners operate in less demanding markets than those controlled by the major oil companies. In such markets older tonnage, often purchased second-hand, is frequently employed. In relational discussions with the head of a Cypriot family that owns one of the largest fleets of oil tankers in the world, he noted an additional criterion that may be added to the four criteria mentioned above regarding standards of operation and maintenance: no ship owner successfully operating second-hand tonnage would sell a vessel as a going concern because it would be used in competition against him. Instead, when a vessel reaches the end of its useful working life a prudent ship owner will sell it for scrap. And because it would be economically illogical to send a ship to the breaker's yard in pristine condition, it does not make economic sense in the final years of a vessel's working life to maintain it to anything other than the basic minimum standard acceptable to flag state, classification society, insurers and charterers.

A major difference between an independent tanker owner and the shipping division of an oil company is that the former relies solely on the vessels to provide an income, whereas the latter has an assured income from the provision of marine transportation to the company's production, refining and sales divisions. Whereas oil companies work out their budgets on a five-year rolling plan and tend to be long-term orientated with regard to their shipping operations, an independent tanker owner may be either long-term orientated or short-term orientated depending on forward contracts.

In exercising due diligence, charterers, particularly major oil companies, increasingly not only demand a high standard of ship operation, especially with regard to safety and environmental protection, they also require chartered-in tonnage to be well maintained structurally, mechanically and also cosmetically. For this very reason the Oil Companies International Marine Forum (OCIMF) set up in the 1990s the ship-vetting programme known as SIRE.

There are undoubtedly ever-increasing pressures on ship owners to raise standards and at the same time to reduce costs, and such pressures are unlikely to change in the foreseeable future. However, while the two goals are at odds with each other, they are the usual pressures found in most organisations, and the pressure to reduce costs does not necessarily result in a lowering of standards.

SUMMARY

This chapter began with an examination of the legal and moral obligations of organisations and individuals at various operational levels in the shipping industry to comply with the ISM Code, and the constraints and pressures that influence the way in which those obligations are addressed.

The legal and operational consequences of non-compliance with the provisions of the Code were also reviewed.

The possibility that non-compliance may be because of economic advantages to be gained by ship owners who avoid implementing the rules was examined. However, it was determined that in general the arguments supporting such a possibility serve only to demonstrate that ship owners are under pressure to cut costs, which is usually the case in any industry supplying a product or a service, and economic considerations are only one of a number of constraints and pressures operating at level 3 of the safety hierarchy.

To complete the literature search and build upon those components of the study reviewed so far, the next chapter explores how personal and societal values relate to safety and risk perception, and the effects of globalisation upon those values.

Subsequently, in Chapter 8, having completed the literature search and reviewed all those culturally and socio-culturally influenced factors that impact upon the development and implementation of an SMS, a dynamic model of the ISM Code is developed.

7

Globalisation

Sailing the Seven Seas

WHY GLOBALISATION IS IMPORTANT

Globalisation is increasingly discussed in cross-cultural management literature for a number of reasons, three of which are germane to the theme of this book.

First, this book is concerned with the impact of the diversity of cultures upon the implementation of common standards of safety within a global industry, and it is important therefore to understand how globalisation is increasingly instrumental in fostering the interplay of cultures by the exchange of values between them.

Second, globalisation means different things to different people and it is therefore important to understand the nature of globalisation in relation to the shipping industry, which operates in a global environment, and to contextualise the constraints and pressures that impact upon the implementation of international Conventions within the industry

Third, the two international Conventions most relevant to the study in this book, the ISM Code and the STCW Code, both relate to safety; one as an organisational tool and the other as an education and training programme. Therefore, in addition to the values and contextual aspects of globalisation, it is necessary to understand how knowledge travels, how safety techniques are disseminated, how objectified ideas[1] such as safety rules, regulations and guidelines may be dis-embedded from one culture and re-embedded in another.

SPREADING OF IDEAS BETWEEN CULTURES

Traditionally, the spreading of ideas was seen as a process of diffusion.[2] However, the diffusion metaphor is limited in its application since it implies that ideas travel in only one direction, cascading from levels of high saturation to levels of low saturation, ignoring situations such as the so-called 'brain drain' where the reverse occurs.

For this reason Czarniawska (see note 1) prefers Latour's translation model,[3] which comprehends the dual association of the word translation: first its association with movement in its meaning of transference, and second its association with language in its meaning of interpretation. The model sees ideas as images which become known in the form of pictures or sounds. The ideas are materialised in the form of objects, which give rise to actions, which may eventually result in the institutionalisation of the ideas.

Czarniawska theorises that an idea can travel only when it has been objectified. A person receiving the objectified idea translates it into a new idea coloured by his or her own values and beliefs (i.e. their culture) and then either discards the idea or objectifies it. And so the process continues, ideas travelling globally and being particularised locally.

Czarniawska's theory applied to two different cultures is depicted graphically in Figure 7 below. In the diagram each idea is translated within a particular culture in localised time and space while the ideas travel from one culture to another through globalised time and space

But ideas do not travel without being affected by various influences. Within each culture and at each level of the safety hierarchy,

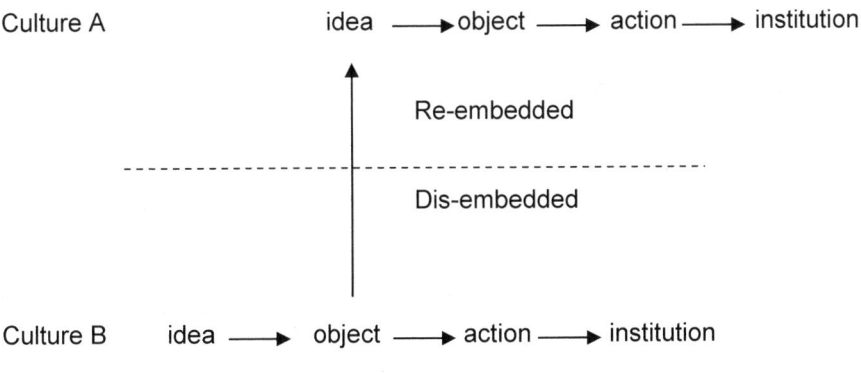

FIGURE 7. How ideas travel.[1]

each idea is subjected to constraints and pressures at each stage of its translation to objectivity, action and institutionalisation. Objects from one train of translation impact upon ideas in other trains of translation and the result will be entirely determined by the interpretation of the receiver.

Moreover, the translation model is not just two dimensional: it is three dimensional with layers of translations occurring horizontally, vertically and laterally as illustrated in Figure 8. Also, Czarniawska's three-dimensional translation model is fractal in two senses. First, as each layer is peeled off then another train of translation is revealed underneath. Second, the greater the magnification of the detail of the influences upon each stage of a train of translation the greater the richness of information discerned.

Hence, close examination of a detailed chain of translation will provide great insight into the prevailing constraints and pressures affecting that chain of translation and the resultant outcome of those constraints and pressures.

RELATIONAL ASPECTS OF GLOBALISATION

Defining globalisation

Amin[4] is of the opinion that the more we read about globalisation from the mounting volume of literature on the topic, the less clear we seem to be about what it means and what it implies: we are assailed by opposing interpretations.

Some writers see globalisation as a new economic world order, organised and run by trans-national financial institutions and international conglomerates. Costea[5] leans towards this concept, as does Mann[6] who referred to globalisation as a hierarchical cosmopolis, i.e. the subjugation of nation states to supra-national, quasi-governmental organisations that promulgate international rules, particularly for the harmonisation of trade and ordering of world peace.

Spybey[7] also leans towards this view, but recognises a cultural aspect to globalisation, stressing the need to retain the view of the individual as a human agent routinely engaged in the reproduction of social institutions, but with the capacity to translate them in the

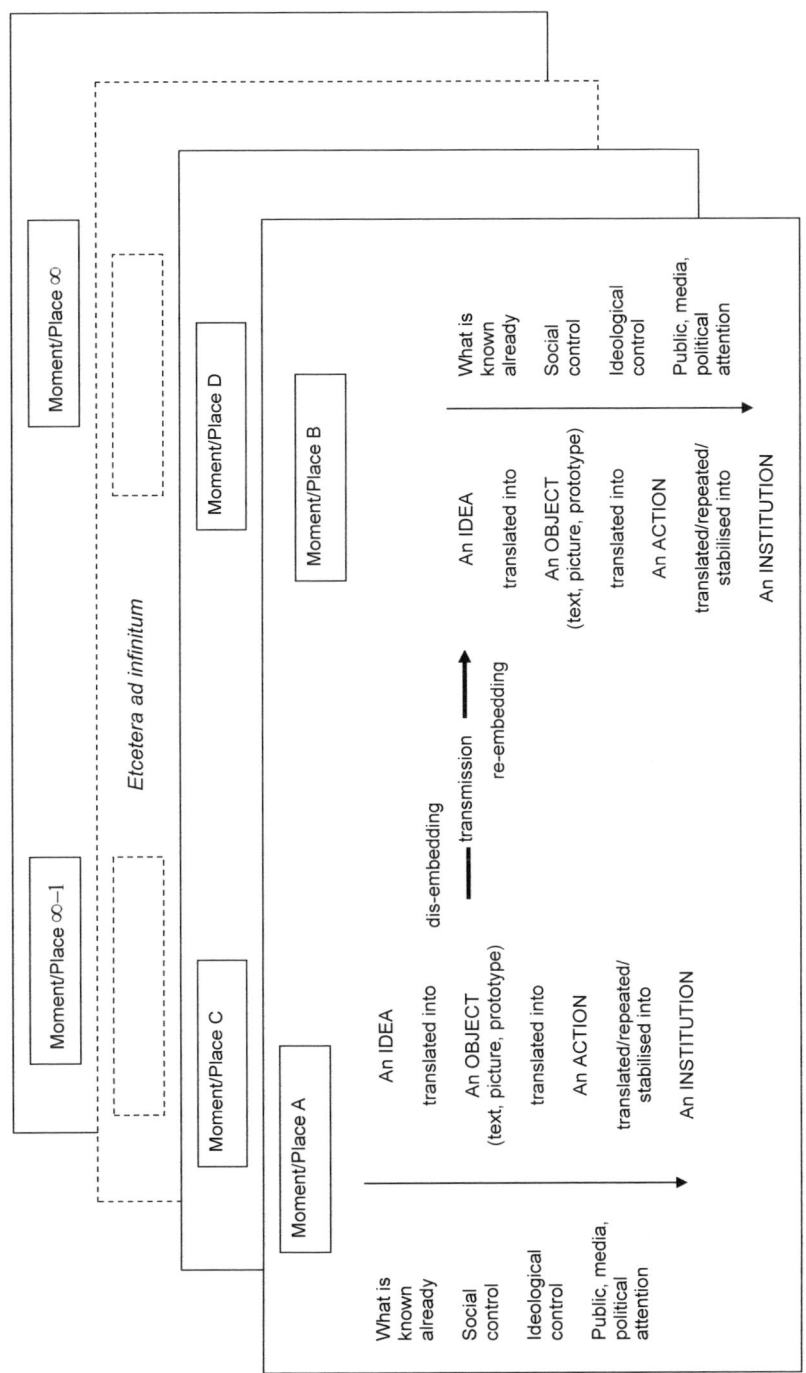

FIGURE 8. Three-dimensional translation model of Czarniawska and Joerges' 'Travels of ideas'.

course of day-to-day activities. This is a view very much in line with that of Czarniawska.

Amin too recognises a cultural aspect to globalisation, but rejects the idea that globalisation is either an economic world order or a hierarchical cosmopolis, arguing that even a cursory awareness of the globalisation literature cannot fail to cast doubt over the idea of globalisation as a neo-liberal conspiracy or simply a system of trade and investment exchange between nations.

According to Amin, even the term neo-liberal has different meanings for different people, depending upon which aspects of globalisation most concern them; whether the economic and political aspects, the socio-cultural and identity aspects or perhaps the very radical concept of global revolution.

As far as socio-cultural aspects are concerned, the neo-liberal conspiracy theory holds that not only is business done internationally in a new way but that there is also a challenge to cultural diversity. For example, the domination of the global media by US companies projects an American view of world affairs, deregulation of international finance leads to a greater division between the rich and the poor in developing nations, and globalisation of large corporations leads to token multi-culturalism at the expense of indigenous non-Western cultures. If this theory is correct then logically the differences between cultures will become less defined as globalisation increases and eventually the entire human race will subscribe to a common set of values and beliefs. This view, however, ignores:

- the difference in styles that exists between systems of corporate governance such as those in Japan, Germany and the United States,[8] which contradicts the neo-liberal idea that globalisation has resulted in there being only one way of doing business in the capitalist world;
- the effects of language. The existence of a diversity of languages tends to support the continuing existence of a diversity of cultures since language is a means of encoding experiences[9] and as Heidegger[10] reminds us, 'Language is the house of being. In this house man dwells';
- localised particularisation of globalised ideas. Ideas interpreted and objictified in one culture are not necessarily coincident with the way in which other people in other cultures interpret and objictify those same ideas.

The extent to which globalisation has blurred the distinctions between cultures has implications for the international agreement and enforcement of maritime safety Conventions such as the ISM Code. Blurring of distinctions between cultures results in more homogeneity in the interpretation of ideas and therefore, a greater tendency towards a common interpretation of the meaning of safety and how safety regulations should be applied. Sharpening of distinctions between cultures on the other hand results in greater heterogeneity in the interpretation of ideas and consequently a greater tendency towards differing interpretations of the meaning of safety and how safety regulations should be applied.

This distinction is important when considering whether greater emphasis on education and training (an attempt at homogeneity) or stricter enforcement of existing regulations (a practice that recognises heterogeneity) is the better path to follow to ensure that the objectives of the ISM and STCW Codes are fulfilled.

Global versus international

The adjectives global and international are often used interchangeably, implying that an increase in globalisation indicates an increase in international trade. But while there may be an increase in globalisation, increases and decreases in the volume of international trade tend to be cyclical. As Amin points out, quoting Hirst and Thompson,[11] at the height of the imperial age between 1878 and 1914 international flows of investment, exports and people exceeded current levels.

Nor is international trade something new. As Hirst and Thompson also pointed out, world capitalism as a system of international exchange is several centuries old. Indeed, international trade and transport have been features of the business world for a very long time: from the era of the Phoenician traders in the Classical Age, to the large international conglomerates of modern times. Domestic producers have always sought ever-wider markets for their goods and are positively encouraged by national governments to export their commodities in order to generate foreign income and thereby increase the wealth of their home nations. Importers have always sought out cheaper or more exotic produce from distant markets to sell in domestic markets. While imports may deplete foreign reserves built up by exporters they also improve the overall material well-being of the home nation.

Importing and exporting are fundamental to the shipping industry but are only a part of international business. Spybey reminds us that the rise of Western states to their position of global economic dominance is inextricably connected with the development of capitalism and the global economy and the extension of Western influence in global terms was coincidental with maritime expansion, the control of international commodity markets and the establishment of an international division of labour.

Whether there is in fact a rational difference between a global and an international organisation or just a conceptual difference is debatable. Perhaps a global company might be distinguished from an international company in so far as the former might imply that the company is not merely trading internationally but is also utilising the personnel, assets and expertise available to it world wide. Lloyd's Register of Shipping, for example, has traditionally been seen as an international company with headquarters in London and overseas offices staffed and managed mainly by British expatriates. However, following recent changes to the way the organisation is structured with more locally recruited staff employed at all levels of operation, both in the UK and overseas, and more autonomy given to regional offices, the organisation may now be seen as a global company

Communications and transportation

Globalisation may in fact simply be business jargon for a world-wide marketplace linked by rapid communications and rapid transportation. Certainly, it is now more fashionable to refer to global markets than to international markets and to talk of globalisation rather than internationalisation. But even if the distinction between international and global described above is valid, both phenomena owe their existence to the advent of rapid communications and rapid transportation.

Developments in communications and transportation have resulted in companies being able to deal in a much wider marketplace than ever before. The advent of air travel in the early twentieth century enabled key staff such as engineers, accountants, and marketing specialists to be sent to overseas offices and factories on an *ad hoc* basis to check up on local operations. The invention of computer-based communications in the late twentieth century enabled overseas offices to report back to head office with consummate ease and speed while the development of container ships introduced a whole new era of sea transportation.

The advent of rapid communications and transportation has also impacted upon international take-overs. According to a newspaper article[12] quoting an OECD report, in 1999 British firms spent £164 billion on overseas acquisitions, accounting for nearly one third of the total amount of money spent world-wide on foreign takeovers.

So it is debatable whether globalisation really exists at all, or whether it is simply a contextual factor emanating from a manifestation of rapid communication and rapid transportation, which can be addressed by conventional approaches to knowledge management and transfer.[13]

Globalisation and culture

With regard to the area of study with which this book is concerned, however, the importance of globalisation is its cultural aspect. Whether it is the cause of, or a product of, technological advances that have precipitated changes to the way that business is administered in the post-industrial age, globalisation may be considered a vehicle of change, a change agent playing an important role in the development of both national and corporate cultures.

Hofstede[14] reasons that we grow from the way in which we respond to our environment and the less an activity is determined by technical necessity the more it is ruled by values, and thus the existence of a global diversity of cultures will produce a variety of different responses to the same idea. However, the way we think, the way we act, our most deeply held beliefs, our very culture, can be changed by the way we perceive and respond to exogenous influences. These changes take place through processes of translation by human agents and this occurs against a background of influences external to immediate social interaction many of which are, in a globalised society, global institutions.

It is self-evident that the ever-increasing speed of communication and transportation exposes individuals, both in nation states and in trans-national corporations, to international cultural influences to an ever-increasing degree. It is the effects of such exposure that are not always clear, for the international traffic of ideas is always a selective process. When considering the implementation of common standards of safety throughout a global industry therefore, it is as noted by Amin essential to look beyond the idea of globalisation as simply a new world order, either economic or organisational, and to look at the impact of globalisation upon regional cultures.

GLOBALISATION, SAFETY AND RISK PERCEPTION

Culture, values and decision making

The cross-fertilisation of cultures must inevitably impinge upon corporate management decision making worldwide. Regional differences will no doubt remain, but the question arises as to whether greater emphasis will be put upon safety and the avoidance of pollution in some countries than in others or whether a general consensus of opinion will arise regarding what is safe and what is unsafe, what is acceptable and what is unacceptable. Even if a general consensus does arise, it remains to be determined whether or not people from some cultures are more likely to take a risk than people from other cultures.

The preferences, values and beliefs of society engender certain expectations. But such preferences, values and beliefs are not static and according to Le Guen[15] current shifts are linked in part to:

- the rapid rise in information technology;
- the increased pace in exploiting advances in scientific and technological knowledge;
- greater affluence in society.

Le Guen's synopsis of the causes of current shifts is closely related to the concept of globalisation as a manifestation of the ever-increasing speed of communication and transportation. He is also of the opinion that these shifts in preferences and values result in:

- a growing perception that risks imposed on people should be justified;
- an increasing reliance by the public on regulators that they trust;

102

- calls for greater openness and involvement in the decision-making process.

With regard to the latter point, Drucker[16] pointed out that management is a decision-making process and whatever a manager does, he or she does through making decisions, whether as a matter of routine or after years of systematic analysis.

But what is important to readers of this book arc the first two points identified by Le Guen. These concern matters that influence managers since they relate to the factors taken into consideration when making decisions. Even strategic decisions are influenced by prevailing environmental and cultural preferences, and safety is, or most definitely should be, a part of the strategy of any well-run company.

Culture, values and globalisation

With the increasing impact of globalisation there is undoubtedly a move towards international standardisation of rules and regulations, and there is evidence to suggest that exposure to the standards and ethics of industrialised nations has influenced newly industrialising nations to follow similar practices, bringing ever-increasing pressure to bear upon national companies, national governments and multinational corporations to adopt socially acceptable and environmentally responsible strategies and safe working practices. For example, in 1997 the local authorities in Nanjing, China, introduced regulations to standardise the region's rapidly growing ship-building and ship repair industry. The regulations require companies wanting to open ship-building or ship repair businesses to ensure that design, construction and reparation of ships comply both with national rules and with international standards.

Similarly, heightened environmental and social awareness in Western societies has put a great deal of pressure on governments and companies in Europe and North America to act in a socially responsible, environmentally aware and safe manner. Governments are pressed by their electorates to demonstrate their commitment to socially acceptable policies by the introduction of public safety regulations and environmentally aware legislation.

Companies too have become increasingly pressed to develop safe and socially responsible operating strategies. When Shell Oil decided that the most economical way to dispose of a redundant North Sea oil rig was to tow it out into the Atlantic Ocean and sink it, the public outcry was so great that the corporation was forced to rethink its strategy. This has since been reflected in the attitude of other major oil companies. Mobil Oil[17] for example, acknowledged that decommissioning offshore production platforms is a controversial subject, presenting operators with both environmental and safety challenges, and recently, as part of a huge project to decommission the North West Hutton platform 130km north-east of the Shetland Islands, British Petroleum undertook extensive consultation with all concerned parties.[18]

However, not all countries have such socially responsible and environmentally aware societies, as can be seen by the lack of social welfare and high levels of pollution in many newly industrialised countries. Safety is also lower on the list of priorities in developing countries, as may be discerned from incidents such as the Union Carbide debacle in Bhopal, India, where thousands of people were killed and injured due to poor practices followed in the local factory of a Western conglomerate.

Managing cultural diversity

Given that there is a global diversity of cultures that impacts in varying ways upon decision making in the fields of safety management, organisational safety and behavioural safety, then as first highlighted in Chapter 1 and reflected upon in subsequent chapters, it remains to be resolved whether the best way to deal with the results of that impact within the shipping industry is by stricter enforcement of existing regulations, such as that undertaken by Port State Control inspectors (a practice which recognises heterogeneity) or greater emphasis on education and training such as that provided for in the 1995 Convention on Standards of Training, Certification and Watchkeeping for Seafarers (an attempt at homogeneity). This question will be returned to and definitively addressed in Part III of the book.

SUMMARY

Globalisation highlights three questions that are at the heart of this book:

1. Are people in some cultures more safety conscious than people in other cultures or are their expectations simply different?
2. Are managers in some cultures more inclined than those in other cultures to put profits before safety, i.e. are they more prepared to 'take a risk'?
3. Can people be educated to the same level of safety consciousness irrespective of their cultural background?

This chapter began by explaining that globalisation:

*   is increasingly instrumental in fostering the interplay of cultures by the exchange of values between cultures; and
*   is an important aspect of this study because it provides the context within which the shipping industry operates and within which IMO is endeavouring to introduce a common standard of safety.

The mechanism by which knowledge travels and safety techniques are disseminated was then explored, and the means by which cultural influences might impact upon ideas as they travel from one culture to another was examined.

An attempt was made to define globalisation. Various writers have expressed a number of different views, defining globalisation in terms such as:

*   a hierarchal cosmopolis;
*   a new economic world order;
*   a neo-liberal conspiracy.

A proposal was then put forward that globalisation might be simply a manifestation of the ever-increasing speed of transportation and communication, providing the context in which trans-national companies operate. International organisations were distinguished from global organisations but it was noted that both benefited from increasingly rapid transportation and communication.

It was noted that greater exposure of cultures to each other could either:

- blur the differences between cultures leading to a greater homogeneity of interpretation of safety regulations; or
- sharpen the distinctions between cultures, leading to greater heterogeneity of interpretation of safety regulations.

This left open the question of whether the best way to ensure effective implementation of the ISM Code objectives is greater emphasis on education and training such as that provided for in the STCW Code, an attempt at homogeneity, or stricter enforcement of existing regulations, an approach that recognises heterogeneity.

Finally, the impact of globalisation upon culture and consequently upon safety and risk perception was examined. It was debated whether that impact presented a challenge to managers or, as Holden contends, merely provided a contextual setting within which global companies operate, the problems being dealt with by conventional approaches to knowledge management and transfer.

This chapter concludes the review of the salient literature, and in the next chapter the themes explored during the review are drawn together to build a dynamic model of the ISM Code. Examination of the model demonstrates that the trains of translation involved in the development and implementation of practices compliant with the ISM Code follow Czarniawska's model of how ideas travel. Further examination of the model reveals that, as outlined in Chapter 6, the nature of the constraints and pressures impacting upon the trains of translation vary in accordance with the level of the safety hierarchy at which they impact and in accordance with the background culture within which the development and implementation of the safety practices take place.

8

A Dynamic Model of the ISM Code

Taking bearings and marking our position on the chart

In this chapter information gleaned from the literature review undertaken in the preceding chapters is used to construct a model of the ISM Code. A similar model could also be constructed of the working of the STCW Code since much of the structure and many of the constraints and pressures are common to both codes. However, such a model would be less useful for present purposes due to a fundamental difference between the ISM and STCW Codes, the difference being that the former is an umbrella code, in so far as compliance with the ISM Code automatically indicates compliance with all other relevant maritime codes. An infringement of the STCW Code therefore would by definition infer non-compliance with the provisions of the ISM Code.

A MULTI-STAGE MODEL

Figure 9 below is a diagrammatic representation of the working of the ISM Code. It is a multi-stage model upon which are super-imposed the five levels of the safety hierarchy developed in Chapter 4, enabling the model to be used as an organising framework to determine where and how cultural pressures and constraints bear upon organisations developing and implementing an SMS in accordance with the ISM Code. The model identifies the various stages at which economic and culturally influenced constraints and pressures might impact upon industry organisations, flag state administrations and hence companies operating vessels under the aegis of those administrations, the entire operation being set within a globalised context.

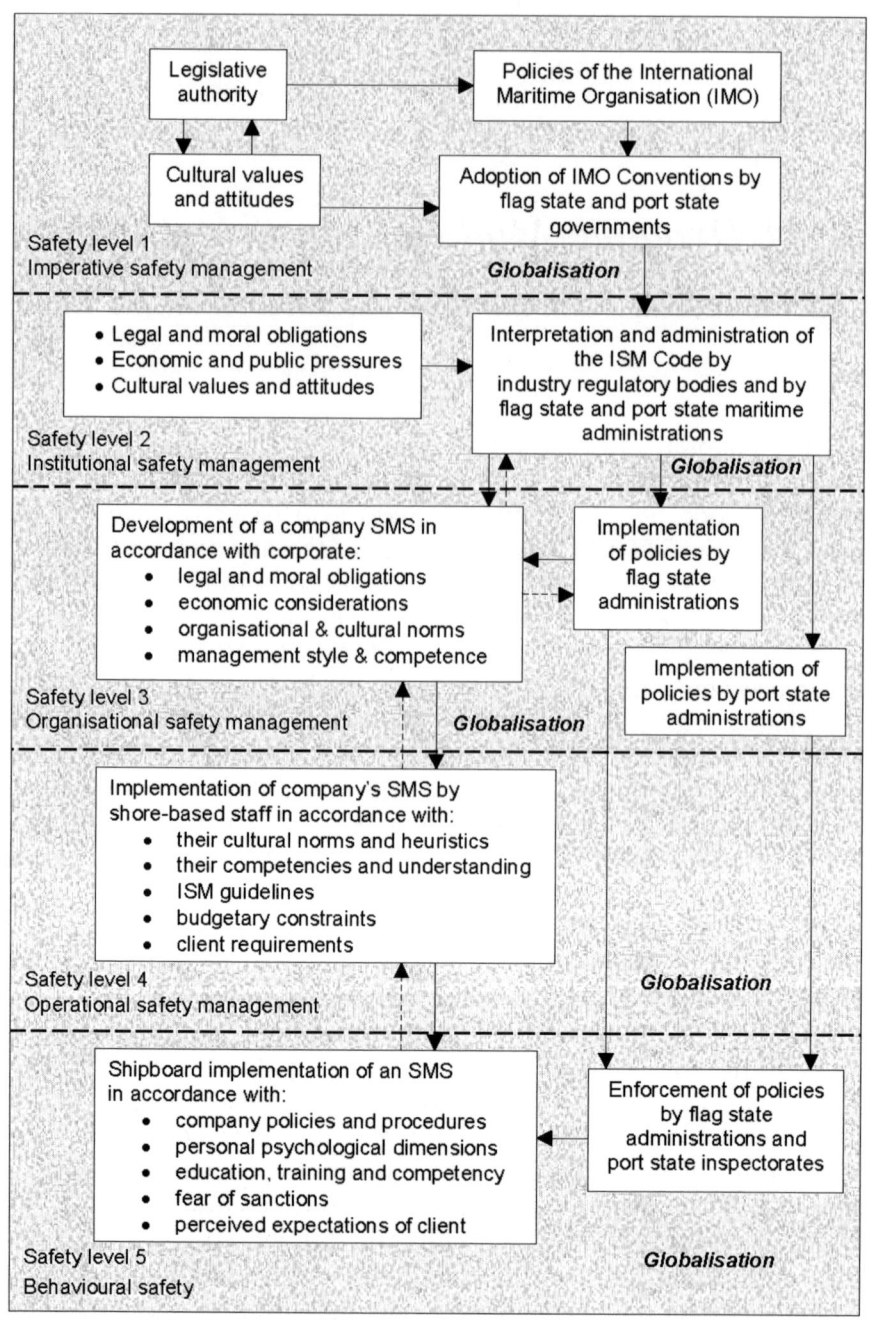

FIGURE 9. ISM Code multi-stage model.

From the figure it is evident that implementation of the provisions of the ISM Code involves a chain of translation, decision making and action. At each hierarchal level of safety management an idea is objectified and travels to the next level of safety management, where it is translated into an idea that is again objectified in a new format and then travels to the next level of safety management. The process is a continuum from the initial adoption of the Convention at an international level and ratification at municipal level by the governments of nation states at safety management level 1, through to its interpretation and promulgation by government agencies and industry bodies at safety management level 2, its adoption as a management tool by senior managers of shipping companies at safety management level 3, its formulation into a system of instructions and procedures by middle management and supervisors at safety level 4, and finally its transformation into positive action by shipboard personnel at level 5 of the safety hierarchy, where it is recognised as the shipboard implementation of an operational safety management system.

The various pressures and constraints that impact upon the development and implementation of an SMS and compliance with the provisions of the ISM Code at each level of safety management are shown in the text boxes in Figure 9 and the entire process is shown taking place in a global context, which in itself presents additional pressures and constraints.

From further examination of the model it is also evident that each hierarchal level of safety forms a superordinate level of management to the one below it and a subordinate level to the one above it. These are not simply different seniority layers but distinct subordinate and superordinate levels of control[1] as discussed in Chapter 4 when comparing SMS models. This is an important distinction from a practical point of view since it helps to clarify the effects of consultancy interventions and information feedback at the various hierarchal levels.

SOCIO-CULTURAL INFLUENCES

Influences at hierarchal level 1

At safety level 1 of the ISM Code model (Figure 9) it is a nation's fundamental socio-cultural and political systems that produce the constraints and pressures which dictate a nation's approach to, and colour its interpretation of, international laws and conventions. The systems comprise political, social and economic institutions, trade unions, social stratification, educational systems and pressure groups. Level 1 of the safety hierarchy, the level at which these constraints and pressures operate, is the level at which nation states send delegations to international forums to discuss and negotiate treaties, conventions and similar concordats. It is the level at which IMO conferences operate.

The constraints and pressures operating at this level of the safety hierarchy were discussed Chapter 6 and may be briefly summarised here as follows:

- The way in which the political and legal institutions have been shaped by the culture and history of individual nations.
- The manner in which a society views law, whether as a body of regulations to be obeyed or as identifiable targets to be achieved.
- The degree to which nation states are industrially developed and are prepared to accept what developed nations consider to be fair and just standards.

Influences at hierarchal level 2

The ISM Code model indicates that at safety management level 2 government agencies, non-governmental organisations and industry bodies such as flag state administrations, port state control inspectorates, and classification societies are either directly responsible, or have assumed responsibility for overseeing the implementation and administration of the decisions made and agreements entered into by nation states.

This represents an intervention at nation state level by the bodies that control the shipping industry. How those bodies respond to and interpret national legislation resulting from national governmental adoption of international agreements will to a greater or lesser degree be influenced by national culture and attitudes towards the implementation of HSE legislation.

Here it should be noted that not only are shipping companies commercially orientated but so too are some of the controlling bodies, particularly the shipping registries of flag state administrations as discussed in Chapter 3 and further evidenced by advertisements regularly placed in the marine press by flag state administrations, such as for example the Dominica Maritime Administration[2] and the Cambodia Ship Registry,[3] in an attempt to persuade ship owners to register their ships with them. And because flag state administrations frequently delegate to classification societies their authority for ship inspections, then by extension the classification societies that vie for such work must also be considered to be commercially orientated. This is not meant to infer any impropriety on behalf of either the flag state administrations or the classification societies but is simply meant to identify where and how commercial pressures operate at this level of the safety hierarchy.

As noted in Chapter 6 and further illustrated in the ISM Code model, the pressures and constraints found at safety hierarchy level 2 are occasioned principally by the same agencies as those at level 1 and may be briefly summarised as follows:

- A legal and moral obligation by maritime administrations of contracting nation states to ensure that the provisions of the ISM Code are enforced.
- Economic obligation by maritime administrations to ensure that ship owners will continue to provide revenue by registering ships with them.
- Public pressure, particularly in countries where a large number of people are engaged in the shipping industry, to remain competitive and ensure that employment within the industry is not jeopardised by inflationary costs.
- Public pressure on maritime administrations to ensure that vessels entering their waters do not cause damage or environmental pollution.

Influences at hierarchal level 3

At hierarchal safety level 3 the ISM Code model reflects the outcome of the interaction between the first two levels of safety management. It is how individual shipping companies respond to the legal and moral obligations brought to bear upon them by flag state administrations, their responses being tailored to the economic circumstances of individual companies, the cultural norms of their management and the country in which their operations offices are established. It is at this level of the safety hierarchy that the empirical research contained in the illustrative case studies in Part II of the book commences with entry into two shipping companies at senior management level.

In the context of Figure 9 the administrative organisations responsible for achieving the aims of the ISM Code, i.e. the provision of an international standard for the safe management and operation of ships and for pollution prevention, lie within both level 3 and level 2 of the safety hierarchy; that is to say within a shipping company's senior management and its flag state administration.

There exists a chain of translation from the level of institutional safety management to the level of organisational safety management that involves interpretation of the purpose of the Code in light of national socio-cultural and political systems prevailing at hierarchal safety level 2 and further interpretation in light of the constraints and pressures prevailing at hierarchal safety level 3.

As noted in Chapter 6 and as illustrated in Figure 9, the principal constraints and pressures that bear upon senior decision makers in shipping organisations may be summarised as follows:

- Both a moral and a legal obligation to run their operations in accordance with the provisions of the ISM Code.
- Consideration of economic factors which may be reflected in budgetary constraints with regard to safety, training and the quality of ship maintenance.
- The prevailing cultural norms of the decision makers; the organisation's corporate culture; management style; and managerial competence.

Influences at hierarchal level 4

Safety level 4 is the level at which middle management and supervisory staff operate. It is here that the policies formulated by senior management are translated into concrete procedures and work instructions to form a corpus of rules that make up the company's safety management system.

As noted in Chapter 6 and illustrated in Figure 9, those middle-level managers and supervisory staff responsible for developing and implementing an SMS will be subject to constraints and pressures generated by:

- the provisions and guidelines of the ISM Code;
- the extent of available resources;
- prevailing organisational and cultural norms, and the heuristics of individual managers and supervisory staff that mould the way in which they interpret the provisions and guidelines of the ISM Code;
- their own professional competencies, proficiency and understanding with regard to the provisions of the ISM Code, safety matters, ship maintenance and ship operations in general.

Influences at hierarchal level 5

Safety at level 5 of the model is safety at the on-board-ship or shop-floor level. It is behavioural safety, safety at the level of the individual, the psychology of human behaviour in relation to the problems of safety in the workplace.

This is the level at which individuals operate ships and have both a moral obligation and a legal duty to do so in accordance with company policies, procedures and work instructions formalised in the company's SMS. Failure to do so may be because:

- the work instructions or safety procedures are ill defined;
- the individuals have a cognitive bias due to a psychological disposition such as an external locus of control which prevents them from taking the initiative when required to do so; or
- the individuals are not sufficiently experienced, well-enough educated or adequately trained to carry out safely the tasks assigned to them.

As summarised in Chapter 5, the principal constraints and pressures acting at this level of the safety hierarchy are:

- A moral obligation and legal duty to follow company policies and procedures.

113

- Educational and training norms as reflected in the competency of individual seafarers. This may also be related to company employment philosophy insofar as that may influence company training policies.
- Those personal psychological dimensions of individual seafarers that might impact upon their behavioural safety patterns.
- A possible fear of sanctions if they do not comply with the provisions of the ISM Code.

AREAS IN WHICH TO CARRY OUT EMPIRICAL RESEARCH

As outlined in Chapter 1, the model of the ISM Code illustrated above in Figure 9 was developed in part as a predictive model and in the main to provide an organising framework showing where and in what manner human factors, especially those that are culturally influenced, might reasonably be expected to impact upon the interpretation and implementation of the provisions of the ISM Code, and with this in mind to determine at which levels of the shipping industry empirical research could most beneficially be carried out.

If subsequent empirical research were to indicate that the diversity of cultures is an influential factor in the uneven interpretation and implementation of the ISM Code, then it falls to be determined whether greater policing by Port State Control inspectors (a practice which recognises heterogeneity) or greater emphasis on education and training (an attempt at homogeneity) would be the better path to follow to ensure that the objectives of the ISM Code are fulfilled.

Taking into consideration the conflicting views discussed in Chapter 1 regarding whether or not the ISM Code is actually effective in achieving its specified objectives, the multi-stage model of the ISM Code may be reviewed to determine the main areas in which the empirical research should be carried out.

From the model, and as noted in Chapter 5, it is at levels 3 and 4 of the safety hierarchy that safety management systems are designed and embedded in the policies and procedures of ship-operating companies, and it is at levels 4 and 5 of the safety hierarchy that those policies and procedures are implemented and education and

114

training are carried out. Therefore it is most likely that empirical research would be most profitably carried out principally at safety levels 3, 4 and 5 of the model of the ISM Code, thus involving:

1. shipping company senior management at safety hierarchy level 3;
2. shipping company middle management, supervisory staff and organisation at safety hierarchy level 4; and
3. sea-going staff at safety hierarchy level 5.

SUMMARY

Development of a model of the ISM Code in this chapter draws together the salient information established from the various strands of enquiry followed in the preceding chapters. These included *inter alia* a history of maritime safety, the nature of safety as a concept, the legal and moral authority of safety regulations, the effects of socio-cultural influences upon human behaviour, aspects of individual psychological dimensions, the impact of globalisation and the importance of establishing effective training policies in individual shipping companies.

Development of the ISM Code model represents an important waypoint in this voyage of discovery being undertaken to study the human factors that impact upon endeavours to raise standards of safety throughout the shipping industry. The waypoint indicates a change of course from the theoretical and academic to the more practical approach adopted in the following section of the book.

In Part II of the book the ISM Code model is used to chart a course for empirical research carried out in two shipping companies. The model is used as an organising framework to determine where and how to measure cultural pressures and constraints that bear upon the organisations in their attempts to develop an SMS in accordance with the provisions of the ISM Code and generally raise the level of operational safety in accordance with the spirit of the Code.

As might be expected, the two companies proved to be rich sources of data and the results of the comparative case study are analysed and commented upon in the concluding section of the book.

PART II

ILLUSTRATIVE CASE STUDIES

9

Preparing for the Studies

Passage Planning

INTRODUCTION

In the preceding part of the book we reviewed the various human characteristics involved in maritime safety and the hierarchical levels at which various associated constraints and pressures prevail, ranging from international governmental level at one end of the hierarchy to the individual behavioural level of seafarers on board ships at the other end of the hierarchy.

The purpose of this part of the book is to bring all those human characteristics and associated constraints and pressures into close focus, illustrating how they might affect the safety performance of a shipping company and how the problems that arise can be dealt with by the company.

It is at this operational level rather than the administrative levels of government and industry bodies that this section of the book is aimed, for it is within individual shipping companies that corporate safety policies and procedures are formulated, safety regulations are promulgated, safety management systems are developed and implemented, employment policies are developed, training programmes are established and work procedures are undertaken. This is the level at which much of the criticism in the marine press regarding the efficacy of the STCW 95 and ISM Codes has been directed.

By way of illustration and to provide a realistic format for an actual consultancy intervention to determine the current status of a shipping company's safety culture, two case studies are presented, based upon an actual comparative case study of two real-world companies carried out by the author to determine the impact of cultural diversity upon the implementation of common standards of safety in the world-wide shipping industry.[1] In order to preserve the anonymity of the two companies concerned they are referred to in this book as Blue Ocean Offshore and Green Sea Offshore.

Both companies are engaged in the ownership and operation of support vessels for the offshore oil and gas industries and have fleets of similar size comprising a variety of vessels such as supply boats, anchor handling tugs, accommodation vessels and safety stand-by vessels. Although each company owns some new tonnage, the average age of the vessels in both fleets is around 20 years, with some vessels exceeding 30 years of age.

The two companies operate in two different geographical regions of the world and employ personnel from countries identified with respect to Hofstede's and Trompenaars' dimensions as being culturally distinct from each other, thus maximising the effects of cultural differences both on the management and on the staff throughout the companies.

The offshore sector of the shipping industry is generally recognised as being particularly hazardous due to the nature of the work involved and has traditionally had a poor safety record in comparison with other sectors of the shipping industry, such as the operation of general cargo vessels or container ships. Consequently, any improvements in operational safety in either company since the introduction of the ISM and STCW 95 Codes could be expected to be more readily identifiable than in companies operating in other sectors of the industry where safety records have traditionally been better.

Blue Ocean Offshore is a financially heavily leveraged company that operates offshore support vessels in the Middle East and South East Asia. All the vessels are managed directly from the company's head office in Dubai with only a Representative Office in Singapore, principally for marketing purposes in South East Asia. The company employs multi-ethnic, multi-cultural shore staff and principally Filipino seafarers.

Green Sea Offshore is a cash-rich company with its head office in eastern England from where it operates offshore support vessels in the North Sea. The company employs entirely British shore staff and almost exclusively British seafarers.

METHODOLOGY AND APPROACH TO THE STUDY

Clause 1.4 of the ISM Code requires companies to develop, implement and maintain an SMS that includes procedures for internal audits and management reviews. In addition, to ensure that ship-operating companies adhere to the IMO rules with regard to the ISM Code, each ship-operating company must undergo an initial flag state audit before it can be issued with a Document of Compliance (DOC) confirming that it has in place an acceptable SMS.

The emphasis on procedures, audits and management reviews infers a system of continuing dialogue between a ship's crew and its Master, the Master and the Designated Person Ashore (DPA) (i.e. the shore-based safety officer who has access to the highest levels of management in accordance with Clause 4 of the ISM Code), and between the DPA and senior management. The emphasis on procedures, audits and management reviews also infers the development of an auditable paper trail, which produces a great deal of both qualitative information such as hazardous incident reports and quantitative data such as accident statistics. Therefore, of the four main research methods available – interpretive, experiment, survey and case study – it is the latter that is most appropriate for the empirical research, a case study strategy being capable of accommodating a mix of both qualitative and quantitative data.[2]

As discussed earlier in this book, there exist three specific elements that are important in the development, implementation and effectiveness of an SMS in a shipping company:

- leadership style / senior management commitment;
- end user involvement; and
- suitability of the SMS.

Associated with these three elements are a number of culturally influenced risk factors that could have either a positive or negative influence on the effectiveness of an SMS. It is therefore necessary to identify and analyse the predominant cultural influences prevailing within each company's organisational structure. This may be achieved by using a number of indicators such as the nationality of senior managers, supervisory staff and vessels' crews, then applying Hofstede's[3] cultural dimensions in order to determine their corresponding positive or negative effects upon the salient organisational elements.

Also, because an individual's locus of control orientation may be influential in determining how useful education and training might be in establishing common attitudes towards safety across culturally diverse groups, Rotter's[4] scale may be used to develop a questionnaire for measuring the locus of control of two samples of seafarers and selected shore staff together with their following details:

- nationality;
- age and experience;
- professional qualifications; and
- academic attainment.

Then, having collected sufficient data from the foregoing procedures, it may be reviewed together with the company's SMS:

- to determine whether or not the SMS is well suited to the organisation; and
- to identify the organisation's strengths and weaknesses that might impact upon the company's overall safety culture.

This enables predictions regarding the effectiveness of the SMS to be made which can be confirmed by reviewing the company's annual safety statistics and safety records together with documentation such as vessel classification status and records of ship detentions as a consequence of Port State Control inspections.

Subsequently, having determined whether prevailing cultural factors have contributed to or detracted from the effectiveness of the SMS, it is necessary to establish the extent to which regulatory control on the one hand and education and training on the other hand are factors in determining the effectiveness of the SMS.

Although a questionnaire style survey could possibly be used for this, it would involve a great deal of subjectivity, both on the part of the respondents and the researcher. Hence, the most cognitively unbiased way of establishing the role both of regulatory control and of education and training is by:

- asking questions by way of formal interview of key personnel; and
- reviewing company documentation, particularly safety records, staff training policies and any references to ship detentions by Port State Control;
- observing the manner in which staff approach safety matters, their degree of involvement and commitment;
- engaging in relational discussions with various company employees both ashore and onboard ship.

From the foregoing it is evident that there are several units or sub-units to be researched. Therefore, although the overall approach to the study is holistic there are also embedded components.

RESEARCH PROTOCOL

To undertake the empirical research effectively it is useful to adopt a protocol that incorporates three phases.

The first phase involves entry into each company at a senior level of management. This is necessary for two reasons. First, as a simple matter of courtesy and practicality it is necessary to gain the permission and cooperation of an organisation's senior managers before embarking upon a study of the company's documentation and interviewing supervisory personnel. Second, as discussed in earlier chapters, commitment by senior management is a *sine qua non* if an SMS is to be effective[5] and therefore entry into each company at a senior level is necessary to facilitate research into the commitment of each organisation's senior managers towards safety management and compliance with the provisions of the STCW 95 and ISM Codes.

The second phase involves research at safety levels 3 and 4 of the ISM Code model and collection of data using the techniques described below. Results from analysis of the data may then be compared and contrasted to highlight how the various constraints and cultural factors have influenced the way in which each company has approached compliance with the STCW 95 and ISM Codes, the type of SMS developed and the degree of effectiveness achieved in satisfying the objectives of the Codes.

The third phase involves research at safety level 5 of the model, the level of behavioural safety. The research uses two means of enquiry, the first of which involves distribution of a two-part questionnaire among representative samples of British and Filipino seafarers to determine the locus of control orientation, rank, experience, and professional and academic qualifications of each respondent. Their locus of control orientation is then compared with the other variables to identify any significant correlations that could help to determine any constraints or pressures resulting from prevailing personality traits or psychological dimensions.

The second means of enquiry in the third phase involves visiting ships, observing their overall condition and holding relational conversations with crew members to gain an impression of how they view the increased emphasis on safety onboard vessels and the ever-increasing amount of paperwork involved.

RESEARCH TECHNIQUES

The study utilises four distinct but complementary research techniques to collect data and, as may be anticipated, there is some degree of overlap,[6] elements of each technique being used in each of the three phases of the study. The four techniques used are:

- formal interview of selected managers and supervisory personnel;
- documentary review;
- questionnaire survey of ship's personnel; and
- relational conversations with individuals and observation of various things such as actions and reactions, management style, incidents arising, etc.

Interviews

Although the ISM Code is prescriptive in outcome rather than in process, certain functional requirements fundamental to its operation are specified, such as the requirements for the company to:

- establish a safety and environmental protection policy (*Cl. 1.4, and 2.1*);
- develop instructions and procedures to ensure implementation of the corporate safety and environmental

protection policy in compliance with relevant international and flag state legislation (*Cl. 1.4.2*);

- ensure that the policy is implemented and maintained at all levels of the organisation, both ship-based and shore-based (*Cl. 2.2*);
- establish training requirements and ensure that it is provided (*Cl. 6.5*).

These requirements are reflected at safety levels 3 and 4 of the ISM Code model illustrated in Figure 9, and from a practical standpoint senior management normally establishes company policy, middle management develops instructions and procedures to implement company policy, and junior management or supervisory staff oversees the implementation of those instructions and procedures.

Based upon these criteria six people at varying levels of seniority, working in the operations head office of each company at levels 3 and 4 of the safety hierarchy, are interviewed regarding their views of their respective company's safety philosophy and safety record, particularly with regard to the implementation and effectiveness of the ISM and STCW 95 Codes.

The objective of this purposive sampling is to end up with a set of key informant interviews[7] that reflect the views of both senior and middle management and also supervisory personnel, thus providing an indication of the prevailing attitude to safety within each company. Detailed profiles of each of the respondents are given in the appropriate section of the next chapter.

A semi-structured interview technique is employed. The selected shore-based staff are interviewed using pre-determined questions drawn up and used as an *aide mémoire* by the researcher to ensure that each respondent is asked the same questions, the directional thrust of which is first to determine the heuristics and biases of the interviewees resulting from their cultural origins, educational attainments and employment backgrounds, and second to discern their commitment to safety in general and also to the successful implementation of an SMS in the company and the establishment of appropriate training programmes.

The list of questions covers:

1. The respondent's personal profile
2. The company profile

3. Vessel reporting procedures
4. Ship manning policies
5. Safety, education and training perceptions
6. Interpretation of the company's SMS
7. Maturity level of the company safety climate
8. Perception of the ISM Code
9. Overall view of safety.

However, the list of questions does not comprise a check-box document given to the respondents to record their responses. The respondents are interviewed individually and the questions are asked in such a manner that there is sufficient latitude for the respondents to answer the questions in their own words. The responses therefore relate to the various pressures and constraints acting upon the individual as shown in the boxes at safety levels 3 and 4 in Figure 9. Finally, transcripts of the responses to the interviews are reviewed, analysed and compared.

For purposes of analysis it would be possible to group the respondents in a number of different ways, such as their level of managerial responsibility, their direct involvement with seafarers, or their own sea-going or shipping company experience, and then to analyse their responses by groupings reflecting those individual aspects.

However, the objectives of the study are to determine:

1. whether significant heuristics or biases emanating from different cultural, educational and employment backgrounds influence the way in which the respondents view safety, particularly with reference to the ISM Code and its implementation;
2. whether education and training such as that stipulated in the STCW Code, or better policing and enforcement of existing regulations is the better path to follow in ensuring the objectives of the ISM Code are achieved.

Consequently, nominal values are assigned to the ethnic and formative backgrounds of the respondents as shown in Table 4. This helps the development of a tabular format for presentation of the interview responses, which in turn facilitates comparison of individual responses relative to a combination of ethnic origins and

126

TABLE 4. Matrix of respondent variables

VARIABLE	CODE	VARIABLE	CODE
North European	1	Tertiary education and structured managerial development	A
Indian	2	Tertiary education and informal managerial development	B
Sri Lankan	3	Secondary education and structured training	C
Filipino	4	Secondary education and no structured training	D

cultural background on the one hand and formal education together with managerial background on the other hand in each of the question areas.

Documentary review

Selected documentation is reviewed to determine the extent to which each company has been successful in implementing an effective SMS within the organisation. The principal documentation utilised for this purpose comprises the company's own safety records, its accident statistics, and the documented SMS.

Reviewing the personnel accident records of each company and comparing them with each other provides one variable against which to measure other variables, such as the cultural dimensions of management and the seafarers employed in each fleet as well as the type of SMS developed by each company.

Comparison of each company's accident statistics with the annual statistics produced by the International Support Vessel Owners' Association for the shipping industry offshore sector as a whole provides a measurement of each company's safety record against the average for the industry sector, thus providing an indication of the overall effectiveness of each company's safety strategy.

Survey of sea-going personnel

Locus of control orientation was identified in Chapter 5 as the most relevant culturally influenced psychological dimension affecting the behavioural safety of individual seafarers. Therefore, as noted above, the third phase of the research includes the development and distribution of a questionnaire designed to measure the locus of control orientation of individual seafarers employed by the two shipping companies.

The questionnaire comprises two parts. The first part consists of questions requiring factual responses regarding the respondent's cultural background, educational attainments and experience. The second part comprises a series of questions based on Rotter's inventory for the measurement of internal–external locus of control orientation.

Blue Ocean Offshore employs mainly Filipino seafarers supplied by crewing agencies in Manila. The use of such agencies is a common practice, particularly among companies employing seafarers from developing nations. Green Sea Offshore employs mainly British seafarers through a crewing agency that is a wholly owned subsidiary of Green Sea Offshore, the sea-going staff being drawn from the general pool of British seafarers. Therefore, questionnaires can be distributed to the seafarers of each company via their respective crewing agency. By this means two sets of data can be obtained from two culturally distinct groups of people, and in order to obviate any linguistic heuristics or biases, the questionnaires distributed in the United Kingdom are written only in English while those distributed in the Philippines are written in both English and Tagalog, the two predominant languages spoken in the Philippines.

Relational discussion and observation

This part of the empirical research involves having objective conversations with people employed at various levels of authority and in various disciplines, and observing how they interact with one another in day-to-day situations and how they act or respond to events in the normal course of their work. The technique involves not only talking to individuals and listening to their comments but also watching their actions, interactions and reactions.

In addition, non-participative observation of organisational actions and reactions to events is undertaken, noting for example how feedback concerning incidents is handled. Also, objective observation of the relative condition of vessels in each company is used to gauge the effectiveness of the company's SMS, in particular the vessel and machinery maintenance systems, and the accuracy of statements made by respondents during interviews and relational discussions.

Written notes are made of conversations and observations considered to be of significance to the study and are used to provide examples, or vignettes, in support of the conclusions drawn from other techniques, thereby helping to counter any tendency to interpret results subjectively.

The conversations and observations run in parallel with the other techniques used during the empirical phase of the study, the objective being to provide data to assist in contextualising the analysis and to assist in determining from an objective standpoint the attitude of senior management, supervisory staff and ships' crews towards the following factors necessary for the development of an organisational safety culture:

- Management commitment and visibility
- Line management commitment and visibility
- Commercial management involvement
- Intercommunication
- Operational pressure versus safety
- Provision of safety resources
- Employment philosophy
- Training programmes
- Mutual trust between management and workforce
- Shared perceptions about safety

Clearly, some of those factors are highly value laden and are therefore more likely to be culturally influenced than others. Identification of the degree to which various factors have been culturally influenced in organisations in culturally divergent countries was one of the main themes of Part I of this book.

SUMMARY OF APPROACH TO THE EMPIRICAL STUDY

Table 5 summarises the research methodology employed for the empirical research, while the next four chapters provide a synopsis of that research, detailing the empirical research carried out into each company's organisational management, operational management

TABLE 5. Summary of research methodology

RESEARCH METHODOLOGY			
Research method	**Comparative case study**		
Research protocol	**Phase I**	**Phase II**	**Phase III**
Safety component	Organisational safety management	Operational safety management	Behavioural safety
Hierarchal safety level	Level 3	Level 4	Level 5
Staff level	Senior management	Middle management and supervisory staff	Seafarers
Research techniques employed	Interview (qualitative analysis)	Interview (qualitative analysis)	Questionnaire (statistical analysis)
	Relational discussion and observation (qualitative analysis)	Relational discussion and observation (qualitative analysis)	Relational discussion and observation (qualitative analysis)
	Documentary review (statistical and qualitative analysis)	Documentary review (statistical and qualitative analysis)	Documentary review (qualitative analysis)

and shipboard behavioural safety respectively, with a view to identifying those data that are commensurate with the aims and objectives of the study, i.e. to:

- identify how human factors such as the diversity of cultures, socio-economic factors, and employment criteria impact upon the implementation of safety regulations;
- determine what obstacles that impact may present to the development of safe practices and attitudes within the shipping industry;
- ascertain whether stricter enforcement of existing regulations or more emphasis on education and training is the better path to follow to overcome those obstacles.

First, Chapter 10 reviews the organisational management of the companies and develops a broad picture of how the two shipping companies respond to their legal and moral obligations, their responses being tailored to the economic circumstances of the individual companies and prevailing cultural norms.

Subsequent analysis in Chapter 11 of the companys' operational management illustrates how individuals respond to the dictates of management, and whether the prevailing cultural norms add additional constraints and pressures at level 4 of the safety hierarchy to those already identified at level 3 of the safety hierarchy.

Then in Chapter 12 the results are presented of statistical and qualitative enquiries into how seafarers respond to safety regulation. Analysis of the results provides insights into whether or not seafarers with different cultural backgrounds respond differently to safety regulations and whether or not education, training and experience are factors that influence their responses.

Finally, Chapter 13 summarises the results of the empirical study presented in the previous three chapters and utilises the information to establish a comparison of the safety culture maturity of the two case study companies. A means of establishing relative safety culture maturity between two companies working in differing cultural environments could be a useful tool in future research.

In Part III of the book the main findings of the study are discussed, conclusions are drawn and areas for further research are identified.

10

Case Study Phase One: Organisational Safety Management

The Voyage of Discovery Begins

ORGANISATIONAL CONSTRAINTS AND PRESSURES

The constraints and pressures prevailing at level 3 of the safety hierarchy and shown on the ISM Code model (Figure 9) were outlined in some detail in Chapter 6. They can be briefly summarised here as follows:

1. Corporate safety culture will influence the type and style of safety management system developed and will reflect:
    a. The organisation's corporate culture;
    b. The prevailing cultural norms of the decision makers;
    c. Management style;
    d. Managerial competence.
2. There is both a moral and a legal obligation to run company operations in accordance with the provisions of the ISM Code.
3. There are economic factors to take into consideration which may be reflected in budgetary constraints with regard to safety, training and the quality of ship maintenance.

The manner in which those constraints and pressures have impacted upon each of the case study companies, and the way in which those companies have dealt with them, are examined in turn in the following sections of this chapter using information acquired from analysis of the key informant interview responses together with information gathered from document reviews, research conversations and observation.

CORPORATE SAFETY CULTURE

Blue Ocean Offshore and Green Sea Offshore developed quite differently from each other and due to their individual histories each company has a distinct corporate culture and each has a safety culture that is different from the other.

Blue Ocean Offshore

Blue Ocean Offshore was formed in the 1990s when a family-owned conglomerate bought the assets of a number of small companies operating offshore support vessels in the Middle East and South East Asia. Although at that time the ISM Code had not entered into force, it had received a lot of coverage in the marine press and senior management instructed operational staff to develop and implement an SMS compliant with the requirements of the ISM Code. The rationale for that decision was not simply to satisfy company safety requirements but also to provide a marketing tool for the commercial department, operational safety being a very attractive commodity to charterers.

Procedures were written, an SMS was duly developed and after a satisfactory audit a Document of Compliance (DOC) was issued on behalf of the relevant maritime administration. However, during the first annual audit a number of serious non-conformities were noted and the DOC was withdrawn. Because compliance with the ISM Code was at that time voluntary the withdrawal of the DOC was of no immediate practical consequence.

However, the formation of Blue Ocean Offshore had taken place during a rising market and was financed by a mixture of private and borrowed capital. Unfortunately the market cooled down and eventually went into a steep decline. The expansion of the company had been quite aggressive in so far as the amount paid for other companies' assets was frequently in excess of their intrinsic value because the purchasers believed that they were buying not just the assets but also market share. Blue Ocean Offshore found itself heavily in debt with a substantial number of poorly maintained vessels and little chance of recovery without substantial inward investment and improved market conditions.

The high level of debt affected not only Blue Ocean Offshore but the whole conglomerate of which it was a part. Consequently, in the late 1990s the conglomerate was taken over by an investment company and a new corporate president, formerly a senior executive in a major international oil company, was appointed. Those changes resulted in financial and corporate restructuring and the recruitment of a new management team selected principally from among former oil company professionals.

Since it is widely accepted that the major oil companies set the standards of safety for oil tankers, refineries and offshore production facilities, it is not surprising that the appointment of a new management team so closely associated with the oil industry brought about major changes in corporate culture and that operational safety was given a high priority, particularly within the Offshore Division.

The high degree of commitment to safety was demonstrated by the decision of the new senior managers at Blue Ocean Offshore to develop without delay a new management system that met the ISM Code requirements in full. A new SMS was subsequently drawn up, selected shore personnel were trained as internal auditors, the new documented procedures were distributed and a successful external audit was conducted following which a new DOC was issued.

Green Sea Offshore

Green Sea Offshore had its roots in 1970. With the demise of the North Sea fishing industry a number of former fishing companies amalgamated and established a company to own and operate offshore safety stand-by vessels in the North Sea. The company was subsequently taken over and re-named Green Sea Offshore in 2000 by a privately owned conglomerate with a wide portfolio of interests. Green Sea Offshore expanded its fleet of offshore support vessels with the purchase of second-hand tonnage. The expansion of the fleet was financed principally out of operating profits and mortgages taken out on the new assets purchased.

A large percentage of the staff of Green Sea Offshore were formerly employed in the fishing industry and subsequently transferred to offshore operations when the initial company was formed. The staff still tended to identify with the company culture established at that time, including its safety culture, rather than identifying with the corporate culture of the new owners, the privately owned conglomerate. Because of this it might be expected that a lower level of safety awareness prevailed in Green Sea Offshore than in Blue Ocean Offshore. However, that was not the case. The responses given during both formal interviews and relational conversations indicated that all Green Sea Offshore personnel were well aware that senior management placed great emphasis on safety and that offshore clients, mostly major oil companies or their subsidiaries, demanded a very high standard of safety to be exercised both by the management ashore and by ships' crews.

Indeed, in both companies the safety culture was not only the result of senior management commitment but was also upwardly driven, with offshore clients demanding ever higher standards of operational safety and a target of zero accidents. Client driven safety requirements applied a constant upward pressure on operations staff and senior management alike, and this is reflected at safety levels 4 and 5 in the ISM Code model (see Figure 9).

Also in common with Blue Ocean Offshore, Green Sea Offshore encouraged the reporting of accidents and hazardous incidents by introducing a no-blame culture in which a person could report a minor accident or a hazardous incident without incurring any blame even though the person reporting the incident was at fault. In neither company did this mean that a seafarer would not be sacked or suffer some form of sanction for causing serious injury, extensive property damage or heavy environmental pollution through negligent or reckless action. In both companies people were expected to take responsibility for their actions. In Green Sea Offshore this was referred to as a 'just culture'.

Because the ISM Code was about to enter into force at the time that Green Sea Offshore came into existence, senior management of the company decided from the outset to establish procedures and implement an SMS that would comply with the ISM Code and hence be acceptable to the British maritime administration.

Lessons to be learned

Upon review of the original SMS devised by Blue Ocean Offshore it was found to have been drawn up in isolation without taking into consideration the actual structure of the company and without employee involvement. It was recognised that the new system would have to avoid similar mistakes and also, because of the diversity of corporate and national cultures inherited from the recently acquired companies, the new SMS would have to be suitable for use by a multi-national management ashore and a predominantly Filipino but intrinsically multi-cultural sea-going staff. These pressures and constraints, together with the fact that the vessels carried small crews, dictated that while the new SMS had to be effective it also had to be kept as simple as possible. That required commitment from senior management and also giving to people involved with the SMS, both ashore and onboard ship, a sense of ownership.

The new senior management of Blue Ocean Offshore recognised that achieving compliance with the ISM Code was only the start of developing a true safety culture within the company and that the introduction of an SMS, whether initially or to replace another SMS, involves an organisational change and requires subsequent monitoring of the system. It is a commonly accepted tenet of change management that to ensure a new system is implemented effectively it is essential for senior management to consistently disconfirm patterns of behaviour not in keeping with the new organisational philosophy, and to consistently support any evidence of movement in the direction of the new assumptions.[1]

In order to raise the safety awareness of individuals, Blue Ocean Offshore employed various tools such as no-blame incident reporting and five-star safety awards for vessels with a good safety record. One tool that stood out as being not only novel, innovative and creative but also practical and successful was the introduction of a quarterly HSE Safety Incidents Booklet, distributed throughout the fleet and also among clients. Not only was the booklet unique in its concept of open reporting of safety-related incidents, but it also had great diversity of content from the statistical to the practical, from the humorous to the serious, reflecting the values of the contributors to the booklet.

In today's environment, which places great emphasis on not only being safe but also being seen to be safe, it is most unusual and counter-intuitive for a company to be so forthright in acknowledging its actual lapses of safety and to advertise them so widely. But this very concept of transparency in accident reporting provided the quarterly HSE Safety Incidents Booklet with a unique and innovative character while also raising the safety awareness of its readers and developing a feeling of trust among the company's customers.

Whilst Blue Ocean Offshore consciously absorbed and developed the strong points of the safety cultures of the companies it had acquired, Green Sea Offshore retained a safety culture that reflected the inherent values that existed prior to the company being taken over by the private conglomerate. However, steady standardisation of company safety culture over a period of years was a long-term objective of both companies, but there could be no short-cuts since in a multi-cultural, global environment it takes time to distil and codify the many different prevailing values and perceptions into a standardised format suitable for developing a true safety culture.

Organisational norms

Because Blue Ocean Offshore was formed by buying the assets of a number of small companies that operated in the offshore sector of the shipping industry, there was no long-established organisational culture associated with the company. The feeling within the head office in Dubai was one of goals and achievements, and even though many of the staff employed in the head office had originally been employees of the small companies from which the assets had been purchased, most identified with their new employer. All Blue Ocean Offshore employees were intent on helping the company to achieve its stated goals and that was quite understandable considering that failure of such a heavily leveraged company to achieve its goals would most probably result in financial ruin, dissolution of the company and employees losing their livelihoods.

As previously noted, Blue Sea Offshore had a multi-national shore management and primarily Filipino seafarers with a sprinkling of Indonesians, Hondurans and East Europeans. Consequently, sound cross-cultural management was a necessity to achieve positive results. The managing director of the company was fortunately well practised in cross-cultural management. A former Master Mariner, he had spent nearly 20 years in shore-based ship management with a major oil company, several of those years in South East Asia and a number in the USA, prior to joining Blue Ocean Offshore.

Green Sea Offshore was the antithesis of Blue Ocean Offshore. Many of the Green Sea Offshore staff tended to identify with their former company (pre-acquisition by private conglomerate) rather than with the new entity that Green Sea Offshore had become. Consequently, the prevailing corporate culture was that of an established company undergoing a process of change management driven from corporate level.

The fact that Green Sea Offshore had an all-British shore staff and employed predominantly British seafarers with a sprinkling of East Europeans lent a degree of homogeneity to the company that facilitated the change management process. This homogeneity was reflected in the similarity of the attitudes of the respondents during formal interviews and relational discussions.

CULTURAL CONTEXT

In terms of Hofstede's cultural dimensions, those prevailing in the geographical regions where the two fleets operated were diametrically opposite to each other in terms of power-distance and individualism/collectivism but similar to each other in terms of uncertainty avoidance and masculinity/femininity. This was also true of the cultural dimensions of the staff employed by the two companies. Utilising the values determined by Hofstede in his study,[2] the index values of the prevailing cultural norms relevant to the

TABLE 6. Matrix of prevailing cultural norms

	British	Filipino	Indian	Arab	Pakistani	Other
Blue Ocean shore staff	9%	6%	70%	12%	–	3%
Green Sea shore staff	100%	–	–	–	–	–
Blue Ocean sea staff	–	47%	30%	2%	6%	15%
Green Sea sea staff	96%	–	–	–	–	4%
PDI score	35	94	77	80	55	–
IDV score	89	32	48	38	14	–
UAI score	35	44	40	68	70	–
MAS score	66	64	56	53	50	–

nationality of the personnel employed by each company are shown in Table 6.

As discussed in Chapter 5, individualism/collectivism and power-distance were determined to be the two cultural dimensions most likely to impact upon organisational management and operational management. The percentages of staff by nationality and the differentials between cultural dimensions by nationality shown in Table 6 for the sea-going and shore-based staff of the case study companies also indicate that those are the two most relevant cultural dimensions in relation to the current case study.

KEY INFORMANT INTERVIEWS

Six people from each company were interviewed regarding their personal views of the company's safety philosophy and safety record, particularly with regard to the implementation and effectiveness of the ISM and STCW Codes, in order to:

- gain a comprehensive view of the company and its organisational structure;
- explore the effects on each company of the legal and moral obligations and corporate economic considerations shown in the ISM Code model (Figure 9) to be acting at Safety Level 3; and
- determine the commitment of each company's managers and supervisory staff towards safety management, employment policies and compliance with the provisions of the ISM and STCW Codes.

THE RESPONDENTS

It was important that the interview responses reflected the views of both senior and middle management and also supervisory personnel, thus providing an indication of the overall prevailing attitude to safety and to the implementation of the STCW 95 and ISM Codes within each company.[3] The respondents were therefore carefully selected using the following two criteria:

1. All had a responsibility within the parameters of their job function for ensuring safe operation of the vessels by the people on board.
2. They represented the various tiers of management and supervision within the companies for which they worked.

On this basis the following respondents were interviewed:

Blue Ocean Offshore	*Green Sea Offshore*
1. Managing Director	General Manager
2. Operations Manager	Operations Manager
3. Operations Superintendent	Technical Director
4. QA-HSE Manager	Safety Manager
5. HR Manager	Personnel Manager
6. Crewing Supervisor	Senior Crewing Coordinator

In both Blue Ocean Offshore and Green Sea Offshore the Safety/HSE Manager was also the Designated Person Ashore (DPA), a role clearly identified in Cl.4 of the ISM Code as providing a link between managers and seafarers.

During the empirical research, Green Sea Offshore did not have an Operations Superintendent, the duties of that position being undertaken by the Technical Director. Therefore, the Operations Superintendent was interviewed in Blue Ocean Offshore and the Technical Director in Green Sea Offshore.

To assist in identifying any heuristics and biases exhibited by the respondents resulting from their cultural origins, educational attainments and employment backgrounds, transcripts of the interview questions and responses were formatted and grouped according to the socio-cultural backgrounds of the interviewees as shown earlier in Table 4, Chapter 9, and are shown in Tables 7 and 8 below and expanded upon in the subsequent text.

TABLE 7. Blue Ocean Offshore key interview respondents

Respondent	MD	Op'ns. Manager	DPA	Op'ns. Super.	HR Manager	Crewing Sup'vr
Ethnic group	1	2	1	4	2	3
Educational/ managerial group	A	B	C	B	D	D

TABLE 8. Green Sea Offshore key interview respondents

Respondent	GM	Op'ns. Manager	DPA	Tech. Director	HR Manager	Senior Crewing Coord'tor
Ethnic group	1	1	1	1	1	1
Educational/ managerial group	A	D	C	B	A	C

North European: 1
Indian: 2
Sri Lankan: 3
Filipino: 4
Tertiary education and structured managerial development: A
Tertiary education and informal managerial development: B
Secondary education and structured training: C
Secondary education and no structured training: D

BLUE OCEAN OFFSHORE RESPONDENTS

Ethnic origins

Two (the MD and the DPA) were British of Anglo-Saxon/Celtic origins.
Two (the Operations Manager and the HR Manager) were North Indian.
One (the Operations Superintendent) was Filipino.
One (the Crewing Supervisor) was Sri Lankan.

Cultural backgrounds

There were a number of differing cultural influences bearing upon the respondents:

141

1. The Managing Director felt that his cultural background was North European although he regarded his country of domicile as UK and UAE.
2. The Operations Manager felt that his cultural background was Indian-Western; English was the language used in his home and he regarded his country of residence as India.
3. The DPA considered his cultural background as Western/Oriental and his country of domicile as Malaysia.
4. The Operations Superintendent considered that he was both ethnically and culturally Filipino and his home was in the Philippines.
5. The HR Manager had lived in the UAE for 25 years together with his family. Such people are known locally as Non-Resident Indians or NRI.
6. The Crewing Supervisor had lived in the UAE for 14 years and although he said his country of domicile was the UAE his family were still living in Sri Lanka.

Commonalities

1. All had worked for the company for between one and five years.
2. All were living in Dubai.
3. Four of the six respondents were ex-seafarers and a fifth had worked on offshore oil rigs and production platforms. The sixth had worked as a shore-based radio operator communicating with vessels.

Occupational and educational background

1. The Managing Director, Operations Manager and Operations Superintendent were all ex-seafarers and each had Class I (Master Foreign Going) Certificates of Competency. The Managing Director also had a Diploma in Surveying from the Nautical Institute and had spent over 20 years in senior positions with a major international oil company.
2. The DPA had completed secondary school but had received no formal tertiary education. His practical safety training and experience, however, enabled him to become a member

of the Institute of Occupational Safety and Health. He had also trained as a Lead Auditor (IRCA).

3. The HR Manager and the Crewing Supervisor had finished secondary school but had no formal academic or professional qualifications.

GREEN SEA OFFSHORE RESPONDENTS

Ethnic origins and cultural backgrounds

All six respondents were British and of Anglo-Saxon/Celtic origins.

Commonalities

1. The Safety Manager and the Personnel Manager had worked for the company for four years and five years respectively. The other four respondents had each worked for the company for between 13 and 19 years.
2. All of the respondents had successfully completed a course in ISM Code Familiarisation & Internal Auditor Training.
3. All were living in England with their families.
4. Two of the respondents were ex-seafarers and a further two had been closely connected with ships and shipping for most of their careers. The other two respondents became involved with shipping only when they joined the company.

Occupational and educational backgrounds

1. After completing his education the General Manager had undergone a management apprenticeship, with systematic exposure to the various aspects of ship operations, maintenance and management in various companies over a number of years, before taking up a position in the newly formed Green Sea Offshore prior to the company's acquisition by the private conglomerate, eventually being promoted to his present position.
2. The Operations Manager started his career working on the fish quays and later joined the shore staff of a trawler company that moved from fishing into the offshore support boat industry.

3. The DPA (Safety Manager) sailed in the catering department of an oil tanker company and later transferred to oil rigs and production platforms as a Camp Boss. The experience gave him a general grounding in HSE with emphasis on hygiene that stimulated his interest in other areas of HSE and he transferred from catering to the safety department on offshore structures, then finally moved ashore into the safety department of Green Sea Offshore.

4. The Technical Director had served at sea as an engineer on cargo ships, cruise ships and offshore supply vessels before taking up a position ashore as a superintendent engineer. He had a First Class Engineer's Certificate of Competency (Steam and Motor).

5. The Personnel Manager had a degree in Business Management majoring in Human Resources, and prior to joining Green Sea Offshore was employed in contracts engineering.

6. The Senior Crewing Supervisor had a background in pensions' administration with an insurance company prior to joining the HR department of Green Sea Offshore some 17 years prior to the interview.

COMPARISON OF INTERVIEW RESPONSES

The tabulated transcripts of the key informant interviews were analysed for comparisons, contrasts and matching patterns. The following summary identifies salient comparisons, contrasts and emergent patterns on a topic by topic basis.

Company profile and vessel reporting procedures

The questions in this sector related essentially to matters of fact rather than matters of opinion and were asked primarily to assist in gaining a comprehensive overall picture of the company structure. Therefore, all respondents' answers were very closely correlated and no significant deviation was noted either across the cultural divide or between companies.

With regard to vessel reporting and maintenance requirements, everyone was agreed that:

- there were condition reports that had to be submitted by vessels to the company on a regular basis;
- there were differing frequencies for different reports;
- the reports were sent to the operations department from where they were distributed to designated departments;
- Vessels were inspected both regularly and also on an *ad hoc* basis, and in addition there was also a requirement for every vessel to be inspected at least once per annum.

All respondents in both companies were aware that their respective company had a planned maintenance system (PMS). In Blue Ocean Offshore all respondents were aware that the PMS was in the process of being changed from a paper-based system to a computerised system. In Green Sea Offshore only the technical department staff appeared to be aware that the PMS comprised two parts: a ship-based system covering basic maintenance and a shore-based system for tracking and monitoring more complex maintenance.

Ship crewing policies

Once again, the questions asked related to matters of fact rather than opinion and there was little in the responses to identify any culturally influenced differences between the interviewees. However, the responses did reveal that the two companies had quite distinct employment philosophies.

Blue Ocean Offshore employed predominantly Filipino seafarers and relied upon three outside crewing agencies to supply most of the seafarers although some were directly engaged. But whether supplied by a third party agency or directly engaged by the company all sea-going personnel were employed on short-term single-voyage contracts.

Green Sea Offshore, on the other hand, employed predominantly British seafarers and engaged them exclusively through a single crewing agency, which was a wholly owned subsidiary of Green Sea Holdings Inc, the parent company of Green Sea Offshore. The personnel employed by Green Sea Offshore had annual rolling contracts of employment with the crewing agency and could therefore be considered as company employees.

The differences in employment philosophies are of some significance when consideration is being given to providing education and training for employees, in so far as companies are more inclined to provide education and training for company employees than for casual labour or short-term employees unless required to do so by regulation.

Safety

Analysis of the responses to the questions regarding safety revealed little overall difference between the respondents across the cultural spectrum. Indeed, there is a great deal of similarity between the responses of the six Green Sea Offshore respondents and two of the Blue Ocean Offshore respondents: the HSE Manager and the Operations Superintendent. It is considered significant that all eight had received ISM familiarisation and lead auditor training.

The views of the other four Blue Ocean Offshore respondents were not far removed from those of the other respondents and it is noted that the Managing Director and the Operations Manager had both received lead auditor training and the HR Manager and Crewing Supervisor had both attended an in-house ISM Code seminar.

Education and training

All respondents were agreed that safety training is effective, although seven of the twelve respondents qualified their agreement with the proviso that the training would be effective only if conducted in a manner that was motivational, provided the trainees with a sense of ownership, was ongoing and had relevant content.

All twelve respondents thought that there was a connection between professional training and safety training. While the Blue Ocean Offshore Managing Director thought the connection was marginal, the other eleven respondents thought the connection was fundamental to safety. The general consensus of opinion was summarised in the response of the Green Sea Offshore Safety Manager who stated, 'Professional training should lead to competence and competence should lead to safety.'

However, despite this consensus it was evident that the training offered by both companies was mainly safety training with only a minimal amount of vocational training or continuing professional development being undertaken. None of the respondents referred to Clause 6.5 of the ISM Code which requires companies to establish

146

and maintain procedures for identifying training requirements and ensuring that the requisite training is provided.

When asked about any changes the company had made to provide for the training requirements stipulated by the revised STCW Code, the responses differed across company lines but not across cultural lines. Blue Ocean Offshore respondents all emphasised checking to ensure that seafarers had the correct, valid and current licences and certificates, while Green Sea Offshore respondents all mentioned training courses provided by the company to ensure that the seafarers they employed received the requisite training. This reflects the different employment philosophies of the companies as outlined above.

Eleven of the twelve respondents thought the ISM Code would be more effective than the STCW Code on raising safety standards in shipping, and the twelfth respondent thought the two codes would have equal effect.

Of the twelve respondents only two in each company were members of a professional body or learnèd society: the HSE/Safety Managers of both companies were members of the Institute of Occupational Safety and Health (IOSH), the Personnel Manager in Green Sea Offshore was a member of the Chartered Institute of Personnel and Development (CIPD) and the Managing Director of Blue Ocean Offshore was a member of both the Nautical Institute (NI) and the Honourable Company of Master Mariners.

Once again, no divisions of opinion were noted along cultural lines in this section of the interview.

ISM system profile

All respondents were aware that their respective companies had developed and implemented safety management systems compliant with the ISM Code well before the required date. It was also pointed out by the HSE Manager of Blue Ocean Offshore that all vessels in the fleet had a valid SMC even though eleven of the vessels (i.e. less than one third of the fleet) actually required an SMC under the regulations because they were below the stipulated minimum tonnage.

In both Blue Ocean Offshore and Green Sea Offshore the HSE/Safety Manager was responsible for arranging both the periodic internal and external audits of the company's safety management system stipulated by the ISM Code provisions.

The questions in this section of the questionnaire related essentially to matters of fact rather than matters of opinion, no culturally influenced differences in the responses were discerned nor any differences of opinion due to varying managerial perspectives.

Safety climate

All respondents in Green Sea Offshore were quite sure that the company had a genuine safety culture and that it was well developed. The respondents in Blue Ocean Offshore were less positive, with the Managing Director stating that the company did not in fact have a genuine safety culture.

The Managing Director of Blue Ocean Offshore and the General Manager of Green Sea Offshore both thought the prime motivation for their respective company's safety management policies was a drive by top management to improve safety performance coupled with a need to comply with regulatory requirements. By and large these views were shared by the other respondents and were reflected in their responses, there being no evidence of any cultural grouping of the views expressed.

All respondents were quite sure that most, if not all accidents and all hazardous occurrences (i.e. 'near misses') were reported. All expressed the opinion that there was no reluctance on the part of ships' staff to report accidents and hazardous occurrences because both companies promoted a no-blame culture, although the Safety Manager of Green Sea Offshore felt that ships' staff tended to perceive the company as having a different safety culture from that which actually existed.

The respondents indicated that both companies had a tiered level of response to accident reports dependent upon the nature and severity of the incident. It was apparent, though, that the accident response in both companies was very much 'top-down' driven rather than of a 'feedback and discussion' nature as often shown in SMS schematic diagrams.

Perceptions of the ISM Code

Respondents were asked whether or not the introduction of the ISM Code had had any appreciable effect on vessel safety, reduction in accidents and improvement in operating standards, and whether or not the ISM Code was achieving its purpose of providing an international standard for the safe management and operation of ships and for pollution prevention. The question elicited varying responses.

Those people closest to the vessels, i.e. those who most frequently go on board for operational reasons or to carry out safety audits, all agreed that they had seen marked improvements and that the ISM Code was achieving both its objectives and its purpose. There was no division across cultural lines. Three of the respondents, however, struck a dissonant note: the Managing Director of Blue Ocean Offshore felt that while the ISM Code was steadily achieving its objectives it had not noticeably made any headway in achieving its purpose of providing an international standard, and the General Manager and the Operations Manager of Green Sea Offshore both noted that the ISM Code was hampered in achieving its purpose because some flag states are not as conscientious as others in enforcing the rules.

Most respondents said that since the introduction of the ISM Code they had seen a significant reduction in accidents or hazardous occurrences and an improvement in the overall condition of the vessels. However, the Managing Director of Blue Ocean Offshore noted that this had coincided with the introduction of other regulations, while the General Manager, Technical Director and Personnel Manager of Green Sea Offshore all felt that the reduction was due primarily to new company policies introduced prior to the introduction of the ISM Code. There was no evident division along cultural lines among the respondents.

The Blue Ocean Offshore respondents reported an improvement in operating standards, but the Green Sea Offshore respondents were less convinced that there had been any improvement in operating standards in their fleet, at least as a result of the introduction of the ISM Code. Again, there was no evident division along cultural lines among the respondents.

In response to the question 'From your experience and in order of priority what do you believe are the three most effective ways of improving overall ship operating standards?' the general consensus of all twelve respondents was:

a.   More emphasis on professional training.
b.   More emphasis on safety training and quality control.
c.   Stricter enforcement of existing regulations.

Two respondents thought that there should be greater regulation of the industry as a whole but when asked how they perceived the degree of regulation in the shipping industry *all* respondents, with the exception of the Blue Ocean Offshore HSE Manager, thought that the industry was already either sufficiently regulated or over regulated.

Overall safety perceptions

Finally, the respondents were asked what they believed were the two most significant potential hazards to safety in the shipping industry. Interestingly, although couched in different terms, one theme was common to all interviewees across both the cultural and managerial divides: a need for greater emphasis on education and training. This can be seen from the responses from the individual respondents.

Blue Ocean Offshore interviewees responded as follows:

- Managing Director: *Different levels of training internationally.*
- Operations Manager: *Insufficient knowledge or commitment from Superintendents.*
- DPA: *Lack of competency.*
- Operations Superintendent: *We have to change the attitude of our seafarers and to do this we have to give them more professional training.*
- HR Manager: *Lack of safety training.*
- Crewing Supervisor: *Low minimum standard of training requirement by the flag state.*

Green Sea Offshore interviewees responded as follows:

- General Manager: *People's competence.*
- Operations Manager: *Shortage of qualified personnel.*
- Technical Director: *Professional competency and lack of basic training.*
- Personnel Manager: *Personnel not properly qualified and trained.*

150

- Senior Crewing Coordinator: *Reluctance of seafarers to change. This could be redressed by more training.*

LEGAL AND MORAL OBLIGATIONS

As noted in Chapter 2 and as further discussed in relation to education and training in Chapter 5, and in relation to the ISM Code in general in Chapter 6, every shipping company has:

- a legal obligation to develop and implement policies and procedures compliant with the provisions of the ISM Code;
- a moral obligation to observe the spirit of the Code and not simply the detail of the regulatory requirements, i.e. it ought to develop policies and procedures that do not simply meet regulatory requirements but also serve to achieve the ISM Code objectives of safety at sea, prevention of human injury or loss of life, and avoidance of damage to the marine environment.

To determine the approach of the management and supervisory staff towards these issues in the two case study companies, each of the respondents was asked what they believed was the prime motivation behind their respective company's safety management policies. The responses to that question produced, perhaps surprisingly, broadly similar responses.

Blue Ocean Offshore interviewees responded as follows:

- Managing Director: *Top management drive it to ensure safety on board the vessels but middle management drive it to comply with regulatory requirements.*
- Operations Manager: *To enhance safety performance.*
- HSE Manager: *To enhance safety performance and in order to comply with safety requirements.*
- Operations Superintendent: *To comply with regulatory requirements, enhance safety performance and avoid legal actions. However, the prime reason is for commercial considerations, e.g. boats under 500 grt do not require ISM but Blue Ocean Offshore do it anyway for marketing reasons – safety is an added bonus.*

- HR Manager: *To comply with regulatory requirements.*
- Crewing Supervisor: *To comply with regulatory requirements.*

Green Sea Offshore interviewees responded as follows:

- General Manager: *50% to comply with regulatory requirements and 50% to enhance safety performance.*
- Operations Manager: *To enhance safety performance.*
- HSE Manager: *Firstly to avoid legal actions. Secondly to comply with regulatory requirements. Thirdly to enhance safety performance.*
- Technical Director: *To enhance safety performance.*
- Personnel Manager: *Primarily to enhance safety performance and as a secondary consideration to avoid legal actions.*
- Senior Crewing Coordinator: *A combination of factors: commercial advantage; regulatory requirements; to keep employees safe; less accidents means less expenditure.*

Analysis of the responses indicates that all respondents in both companies were not only well aware of the legal requirements imposed upon them by the introduction of the ISM Code but also fully supportive both of its introduction and of the need for the companies to implement its provisions.

The similarity of views expressed across the cultural divide may have been due in part to the effects of globalisation blurring cultural differences, but most probably in the main to all respondents having undergone ISM Code training. Eight of the twelve key informants had attended identical training courses covering ISM familiarisation and auditor training, two had attended in-house ISM introductory seminars and the remaining two were qualified lead auditors of management systems. This is supportive of the argument for the introduction of common standards of training to achieve a common standard of safety in a global industry.

With regard to the moral obligations imposed upon the company and its employees to act within the spirit of the ISM Code and not merely in strict accordance with its regulatory provisions, all respondents were very positive about the potential benefits of the introduction of the ISM Code, although in the case of the Blue

Ocean Offshore Operations Manager it was felt that his responses might have been influenced to some extent by social desirability.

ECONOMIC CONSIDERATIONS

When the new owners took control of Blue Ocean Offshore the company was very heavily leveraged with a debt of some US$800 million. The need to reduce that debt was a critical factor in the decisions of the owners and was reflected in the decision making of the company's senior executive managers. Strategic steps were taken either to reverse losses by applying standard management techniques or to cut losses by retreating from particular areas of operation as illustrated by the examples in Table 9.

However, economic considerations were not allowed to impact upon safety in operational areas in either Blue Sea Offshore or Green Sea Offshore. From an examination of each company's records, analysis of the responses to the relevant interview questions, relational conversations and observation it was possible to establish that areas of safety potentially subject to cutbacks as a result of economic considerations were in fact not affected by cost containment policies.

It was clear that both companies sought economies wherever and whenever money had to be spent. Shipyards were invited to tender for repairs, purchase orders for spare parts and consumable items were placed on the most economical suppliers, bulk order discounts were sought, staff headcount was reduced where possible and

TABLE 9. Examples of Blue Ocean cost containment measures

Problem area	Resolution
Company ship repair yard sustaining heavy losses	Ship repair yard sold
High overheads	Office relocated and staff numbers reduced
Cost of stockpiling equipment remaining from former project	High-value parts distributed to vessels. Other parts and equipment sold. Warehouse closed
Large outstanding payments receivable	1. Specific staff assigned to debt recovery 2. Future charters covered by letters of credit

TABLE 10. Operational Effects of Economic Constraints

Area of potential cost saving impact	Potential impact	Blue Ocean observed status	Green Sea observed status
Class surveys	Delayed	Up to date	Up to date
Flag state surveys	Delayed	Up to date	Up to date
Voyage repairs	Postponed	As required	As required
Dry-dockings	Postponed	Up to date	Up to date
Vessel appearance	Corrosion	Well painted	Well painted
Safety audits	Postponed	Done when due	Done when due
Safety training	Not done	As required	As required
Safety equipment	Bare minimum	Above minimum	Above minimum
Spare parts	Delayed delivery	Prompt delivery	Prompt delivery
Consumable items	Not supplied	As requested	As requested

insurance requirements were regularly reviewed with a view to reducing premiums. However, those measures were simply good commercial practice and as is illustrated in Table 10 had no adverse impact upon matters of safety.

Relational conversations did reveal one area relating to corporate and economic governance where both companies were under pressure to observe certain strictures imposed by American law. Following the introduction of the Sarbanes Oxley legislation in the USA in 2004 as a consequence of large corporate financial scandals involving Enron, WorldCom, Global Crossing and Arthur Andersen, all publicly traded US companies and non-US companies with a US presence are required to submit an annual report of the effectiveness of their internal accounting controls to the Securities Exchange Committee. Since the conglomerates that owned Blue Ocean Offshore and Green Sea Offshore both had a US presence and were financed to a greater or lesser extent by American financial institutions, both of the companies were required to comply with the provisions of the Act, which is concerned with corporate governance and increased financial disclosure.

MANAGEMENT STYLE AND COMPETENCE

In Chapter 6, three important points were developed:

- It was noted that of the four risk factors considered important in the development and implementation of an SMS, two factors, leadership style and end user involvement, are also particularly subject to the influence of power-distance and individualism/collectivism.
- The importance of leadership style and the direction it gives to organisational culture was discussed and it was argued that without senior management commitment there will be only token commitment to safety within any organisation.[4]
- It was further argued that unless end users of a safety system are given a sense of ownership through participation and involvement in the development and operation of the system, then safety procedures may not be followed with any degree of enthusiasm, if at all, and in order to achieve corporate safety goals[5] authoritarian measures may need to be employed.

In a high power-distance culture there is a high degree of stratification and each layer of management distances itself from the layer below, leading potentially to management by command rather than management by consultation. This does not lend itself to feedback analysis nor to a climate of end user involvement or empowerment. A high power-distance index therefore tends to increase the risk of lack of end user involvement or commitment, whereas a lower power-distance index tends to have a positive effect and decrease the risk.

From the matrix of prevailing cultural norms shown earlier in Table 6 it can be seen that Green Sea Offshore was not challenged with regard to national culture. Blue Ocean Offshore, however, was located and operating in an environment where the prevailing cultural norms of the local people and of the staff employed by the company exhibited a high power-distance index and a strong collectivist tendency.

Power-distance and collectivism were identified in the literature review as being the two cultural dimensions having the greatest influence upon leadership style and hence end user involvement. The negative tendencies of those prevailing cultural norms are autocratic management with a rigid hierarchy and an organisational culture wherein subordinates do not question their superiors, and managers do not get involved in matters they have delegated to their staff.

From relational conversations with senior managers it was evident that such a leadership style and corporate culture had in fact existed in Blue Ocean Offshore under the previous management. That was a major reason why, as discussed earlier, the original SMS was ineffective resulting in the company's original DOC being withdrawn. However, from the relational conversations and the personnel profiles obtained during the key informant interviews it was also apparent that Blue Ocean Offshore's new managing director and new HSE manager both had backgrounds that prepared them for cross-cultural management. Both men had:

- spent several years living and working in Asia and the Middle East;
- spent their working lives in a marine and/or offshore environment;
- always worked in a multi-cultural environment.

Because of their backgrounds both men would have been well aware of the potential negative effects of the prevailing cultural dimensions and would have been instinctively prompted to take appropriate steps to ensure that the potential pitfalls were avoided. Indeed, the effectiveness of their cross-cultural management skills was evidenced by the fact that in less than two years they had been able to develop and implement an entirely new and ultimately effective SMS in a company with a multi-ethnic, multi-cultural workforce.

The twelve key informants were requested to complete the questionnaire used in the survey to measure the locus of control orientation of seafarers using Rotter's inventory and scale as described in Chapter 5 and later in Chapter 12. The results of the key informants' responses are shown in Table 11. A high score indicates a strong external locus of control orientation and a low score indicates a strong internal locus of control orientation.

Although the sample size was too small to draw any firm conclusions, the results are interesting in so far as they do indicate that there may be a correlation between job function and locus of control

TABLE 11. Key informants' locus of control orientation

	Blue Ocean Offshore LOC orientation	Green Sea Offshore LOC orientation
MD/GM	52.2%	43.5%
Operations Manager	*	43.5%
HSE Manager	14.3%	17.4%
Ops Super/Technical Director	21.7%	39.1%
HR/Personnel Mgr	13.0%	13.0%
Crewing Supervisor/ Coordinator	30.4%	56.5%

* Declined to take part in the survey

orientation that is unaffected by cultural differences. That is an area which may benefit from further research.

SUMMARY

This chapter examined in turn each of the constraints and pressures identified at safety level 3 of the ISM Code model (Figure 9) as those potentially influencing the management decisions of the companies being studied.

From the analysis of the key informant interviews and the overall content of this chapter it is clear that there was a genuine commitment by the senior management of both case study companies to establish a genuine safety culture, while the similarity of respondents' views regarding safety reflected the influence of the similarity of the training undertaken by the respondents.

It is also evident that the management teams of both companies recognised and understood their legal and moral obligations with regard to safety, and in neither company were economic considerations allowed to dictate safety policy.

What comes out very strongly from this part of the case study is that constraints and pressures resulting from legal and moral obligations, economic considerations and organisational and cultural norms not only impact upon organisational safety management but can also be suitably dealt with by good strategic and organisational management using standard management techniques which in a culturally homogeneous company may be of an indigenous nature but in a culturally heterogeneous company require the addition of well-developed cross-cultural management skills.

Whether there is a correlation between locus of control and job function irrespective of cultural background was identified as an area where future management research might usefully be undertaken.

11

Case Study Phase Two:
Operational Safety Management

Full Ahead on Passage

The second phase of the case study research involved a documentary review within each company. The four main areas of documentary review were:

- The company's documented safety management system
- Accident reporting and follow-up procedures
- Accident statistics
- Training policies and procedures.

SAFETY MANAGEMENT SYSTEMS

Both Blue Ocean Offshore and Green Sea Offshore had developed a closed cycle SMS as described in Chapter 4. Considering their completely separate development the systems were remarkably similar, possibly because both companies were engaged in the same sector of the shipping industry.

The SMS of both companies comprised a comprehensive documented system in a tiered format under three heads:

- Company Policies
- Company Procedures
- Company Forms

Each of the three heads was further sub-divided so that specific areas could be referenced with comparative ease.

The comprehensive documented SMS of each company was available in both computerised and hard-copy format. There was some debate among sea-going staff in both companies about which was the preferred format and the arguments tended to hinge not on a person's computer literacy as one might suppose but on their job function. Chief Engineers tended to prefer the computerised format simply because it took up less space, which is a limited commodity on offshore support vessels, whilst Masters tended to prefer the hard-copy format because it was easier to reference than a computerised version while alone in the wheelhouse handling the vessel.

Since in both companies the official language of communication was English, the SMS in each company was in English. Each SMS had controlled and uncontrolled copies, the controlled copies being updated on a regular basis, document control in each company being the responsibility of the safety department.

PLANNED MAINTENANCE SYSTEMS

Although the ISM Code does not specifically require ship-operating companies to develop and implement a planned maintenance system (PMS), the Code does require them to have in place a system for ensuring proper maintenance of the ships and their equipment. Many operators find that a PMS is the easiest way to address this particular provision of the Code and therefore, for all practical purposes, the PMS forms part of the SMS.

However, each case study company had in place a self-contained PMS that could be reviewed separately from the rest of the SMS. Each company's PMS was completely different from that of the other but both required a great deal of back-up from their respective company's technical department.

Blue Ocean Offshore PMS

Some months before the study took place in Blue Ocean Offshore, the company had changed from a paper-based to a computer-based PMS. The computer program used a Microsoft Windows operating platform and the system was designed to be as straightforward as possible while covering routine maintenance of all shipboard machinery except for major machinery overhaul, the latter being

160

scheduled by shore-based technical staff to be carried out by contractors.

In addition to the computer hardware and software the system comprised three manuals:

1. Administrators Guide with program installation instructions (14 pages).
2. User's Manual (25 pages) with instructions covering *inter alia:*
    –   Computer program operating instructions
    –   Data transfer and back-up systems
    –   PMS description
    –   PMS schedule
    –   Job cards for each machinery item
    –   Vessel survey status
    –   Ship information system
3. Defect System User's Guide (nine pages) designed to record and manage defects that occur during operation of the vessel and which are not covered in the planned maintenance system.

Since a large percentage of the Filipino sea-going staff was not computer literate, implementation of the system was certainly not straightforward and a great deal of back-up from the company's shore-based technical staff was required before even a semblance of operational success was evident. That was not due to cultural factors: senior managers had simply overestimated the educational standards and computer literacy of the Filipino seafarers.

Green Sea Offshore PMS

The PMS used by Green Sea Offshore was designed to be as simple and easy to operate as possible. The paper-based system comprised:

1. An Engine Room Maintenance Sheet listing some 105 items, each with a unique reference number.
2. The items were divided into sections showing which items had to be carried out daily, weekly, fortnightly and monthly.

3. Each page of the Engine Room Log Book had on it four maintenance related sections:
   – One listing the reference numbers of the daily maintenance items with space for a signature against each item when completed
   – One for noting any maintenance or repairs carried out
   – One for spare gear used (including filters)
   – One for recording the running hours of the principal machinery: main engines, generators, bow thruster engine and fire pumps
4. A summary page at the back of the log book with two maintenance related sections:
   – One listing the reference numbers of all 105 maintenance items with space for a signature against each item when completed
   – One listing all main machinery with an adjacent space against each item for comments and a check box to indicate its condition (A – Good running order, B – Running but needs maintenance, C – Not running awaiting opportunity to repair, D – Not running require assistance to repair).

Mechanical maintenance of principal machinery which had to be undertaken on a regular or periodic basis was scheduled by shore-based staff in the company's technical department and carried out either in the company's own workshops or by independent contractors. The work was carried out during port turnarounds whenever possible or during scheduled dry-dockings when necessary.

Only simple maintenance tasks were entrusted to sea-going staff. Green Sea Offshore's PMS placed the main onus for repair and maintenance, both electrical and mechanical, on the shore-based technical staff. This was not a culturally related factor but indicated either:

- a lack of trust by senior shore management in the ability or conscientiousness of the sea-going staff; or
- continuance of a system introduced when the company operated fishing vessels on board which the sea-going staff was of a much lower technical calibre.

Socio-cultural inferences

Vessels operating in the offshore sector of the shipping industry are usually crewed by seafarers with Certificates of Competency that are limited by horsepower for Engineers and by tonnage for Masters and Mates. Such certificates are of a lower academic and professional standard than those held by officers on vessels employed in the deep sea trades. This fact together with the small number of personnel on each vessel indicates that the decision by both companies to implement a closed cycle SMS was therefore appropriate.

It would also not be surprising to find that a company operating vessels in the offshore sector of the industry had a PMS that was somewhat basic in nature such as that used by Green Sea Offshore. However, with the growing familiarity that people have with computers and the wide availability of computer training courses in the United Kingdom, one might have expected the introduction of a simple computerised system in Green Sea Offshore.

What was surprising, however, was that Blue Ocean Offshore had implemented a relatively sophisticated PMS even though many of the Filipino seafarers were not computer literate. The company took steps to address that problem, first by advising crewing agencies that officers they supplied must be computer literate, and second by providing senior sea-going personnel with the requisite computer training where appropriate. It was evident that the training was effective because in a short space of time the computerised PMS was fully operational.

There do not appear to be any cultural inferences to be drawn from this, only a difference of approach by the senior management of the two case study companies. What is significant, however, is that the vocational training provided by Blue Ocean Offshore to overcome a specific problem was successful, which lends weight to the argument for greater emphasis on education and training rather than stricter enforcement of regulations in order to improve operational safety standards.

163

ACCIDENT REPORTING AND FOLLOW-UP PROCEDURES

In both companies accidents and hazardous occurrences were immediately reported by the vessel to the DPA, by means of the appropriate report form in the case of minor incidents and by means of telecommunications followed up with the appropriate report form in the case of major incidents.

The reports included the Master's or ship-board Safety Officer's evaluation of the cause of the incident, extent of the damage, pollution or injury, and steps taken or to be taken to redress the problem and to avoid a recurrence of the incident. If the incident was of a serious nature a safety alert would be sent out to all vessels in the fleet to avoid a similar accident occurring on another vessel. Meanwhile, the incident would be included in the DPA's weekly (Green Sea Offshore) or monthly (Blue Ocean Offshore) safety report, which was sent to the General Manager and Managing Director respectively and copied to all vessels.

By these means, important experiences were shared immediately they happened and less important experiences were shared soon after they occurred.

RECORDING ACCIDENT STATISTICS

As noted in Chapter 5, the most common method of evaluating the effectiveness of an organisation's SMS is the recording of accidents, lost time incidents and hazardous occurrences (near misses).[1] But when using such data to measure the effectiveness of a company's SMS or as a means of comparing a company's safety performance with that of other companies, it is important that each company uses the same criteria for categorising accidents and the same mathematical procedures for preparing the datasets used in calculating the statistics that are to be presented.

The Marine Injuries Reporting Guidelines[2] developed by the Oil Companies International Marine Forum (OCIMF) and used by Blue Ocean Offshore are widely used throughout the shipping industry, particularly those sections of the industry associated with the production, storage and carriage of crude oil and petroleum products.

The Occupational Safety & Health (OSHA) guidelines used by Green Sea Offshore are very similar to the OCIMF guidelines so direct comparisons between the accident statistics of the two companies could readily be made.

The two principal variables most commonly measured on an ongoing basis and presented as a monthly or annual statistic are Lost Time Injury Frequency (LTIF) and Total Recordable Case Frequency (TRCF). Each of these figures is a composite, calculation of which is carried out using the following definitions and formulae in line with OCIMF and OSHA guidelines.

Definitions

- *Incident:* An event that results in a fatality or injury to a seafarer onboard ship or while ashore on company business.
- *Lost workday case (LWC):* An injury that results in an individual being able to carry out his or her duties or return to work on a scheduled shift on the day following the injury.
- *Restricted work case (RWC):* An injury that results in an individual being unable to perform all normally assigned work functions during a scheduled shift or being assigned to another job on a temporary or permanent basis on the day following the injury.
- *First aid case (FAC):* Any one-time treatment and subsequent observation or minor injuries such as bruises, scratches, cuts, burns, splinters, etc.
- *Medical treatment case (MTC):* Any work-related loss of consciousness, injury or illness requiring more than first aid treatment by a qualified medical practitioner.
- *Permanent partial disability (PPD):* Any work injury that results in the complete loss, or permanent loss of use, of any member or part of the body that partially restricts or limits an employee's ability to work on a permanent basis at sea.
- *Permanent total disability (PTD):* Any work injury that incapacitates an employee permanently and results in termination of employment on medical grounds.
- *Exposure hours:* Total number of hours a vessel's crew is exposed to the possibility of suffering the consequences of

165

an incident (24 hours per person per day while serving on board).

Formulae

Utilising the above outline definitions, the following formulae are used for calculating accident statistics.

Lost time injuries (LTIs) are the sum of fatalities, permanent total disabilities, permanent partial disabilities and lost workday cases, or:

$$\text{LTIs} = \text{Fatalities} + \text{PTD} + \text{PPD} + \text{LWC}$$

Total recordable cases (TRCs) are the sum of all work-related fatalities, lost time injuries, restricted work injuries and medical treatment injuries, or:

$$\text{TRCs} = \text{LTIs} + \text{RWCs} + \text{MTCs}$$

Each of these indicators is turned into a frequency rate by dividing the indicator by the exposure hours and multiplying by the factor in terms of which the frequency is to be expressed. For example:

$$\text{LTIF} = \text{LTIs} \times \frac{200{,}000}{\text{Exposure hours}}$$

This will give a lost time injury frequency expressed as the number of lost time injuries per 200,000 man-hours. Similarly:

$$\text{TRCF} = (\text{LTIs} + \text{RWCs} + \text{MTCs}) \times \frac{200{,}000}{\text{Exposure hours}}$$

This will provide a total recordable case frequency expressed as the number of total recordable cases per 200,000 man-hours.

The frequencies may be expressed as a rate per any desired unit of exposure hours, but 200,000 is commonly used in the United States and was the unit used by both Blue Ocean Offshore and Green Sea Offshore.

COMPARISON OF ACCIDENT STATISTICS

Accident statistics are by definition quantitative in nature and therefore present a convenient method of assessing the effectiveness of a company's SMS. Trends can be analysed, comparisons can be made with other companies and benchmarking can be achieved by comparing a company's statistics with those of the industry sector overall. Charts of the accident statistics per 200,000 exposure hours for a three-year period in Blue Ocean Offshore and in Green Sea Offshore are given in Figures 10 and 11. The different styles of graphical presentation were those used by the companies and serve to emphasise the fact that provided the data are gathered and statistics calculated using the same guidelines then direct comparisons between the accident rates can be readily made regardless of the presentation format.

BLUE OCEAN OFFSHORE
Personnal incident rates – 36 months

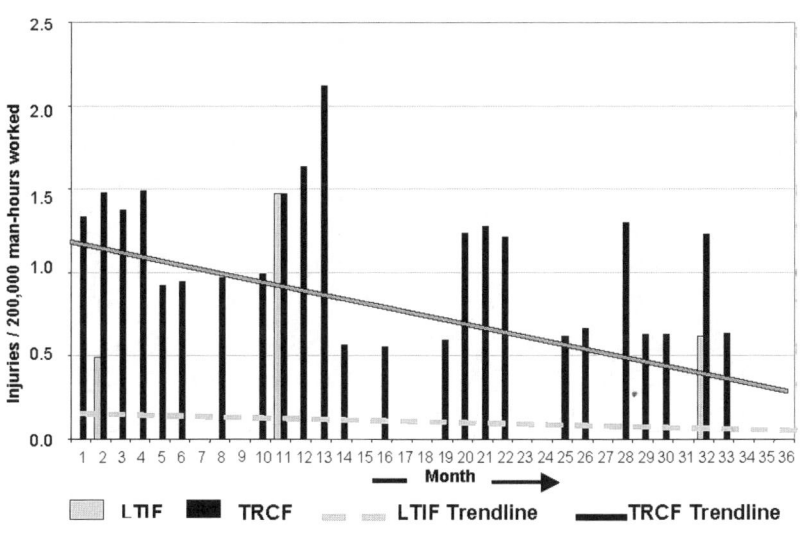

LTIF – Lost Time Incident Frequency   TRCF – Total Recordable Case Frequency

FIGURE 10. Personnel accident statistics – January 2002 to December 2004.

GREEN SEA OFFSHORE
Personnal incident rates – 42 months

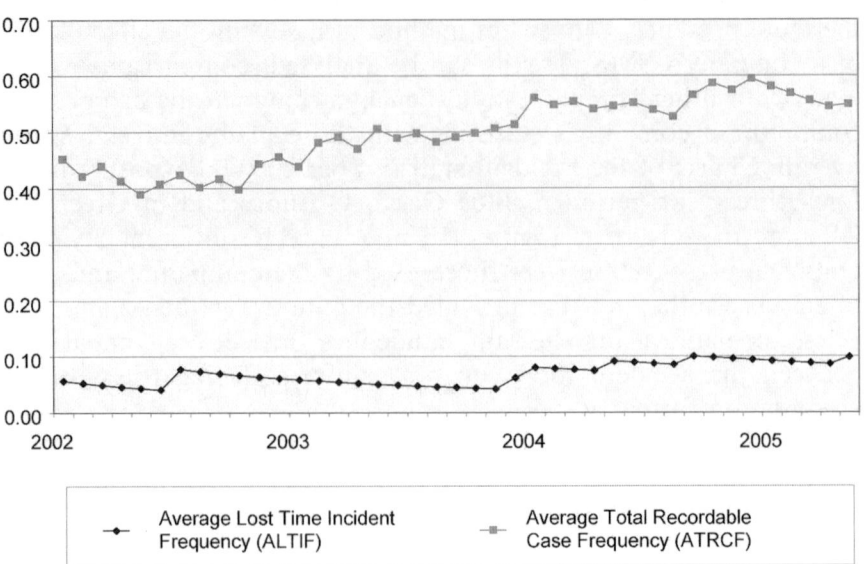

FIGURE 11. Personnel accident statistics – January 2002 to June 2005.

From Figures 10 and 11 it is quite evident that during the periods covered:

- Blue Ocean Offshore has been extremely successful in lowering the frequency of both lost time incidents (LTIs) and total recordable cases (TRCs) with zero LTIs and TRCs in some months;
- Green Sea Offshore has maintained a low frequency of LTIs but has experienced a rise in the frequency of TRCs; and
- by 2004 both companies had achieved similar LTI frequencies but Green Sea Offshore had a significantly higher TRC frequency than Blue Ocean Offshore.

Figure 12 shows the overall lost time accident frequency rates for the offshore sector of the shipping industry as recorded by the International Support Vessel Owners' Association (ISOA) for the years 1998 to 2003 inclusive. The data were extracted from the Associations' annual Personnel Accident Survey 2003[3] and converted so as

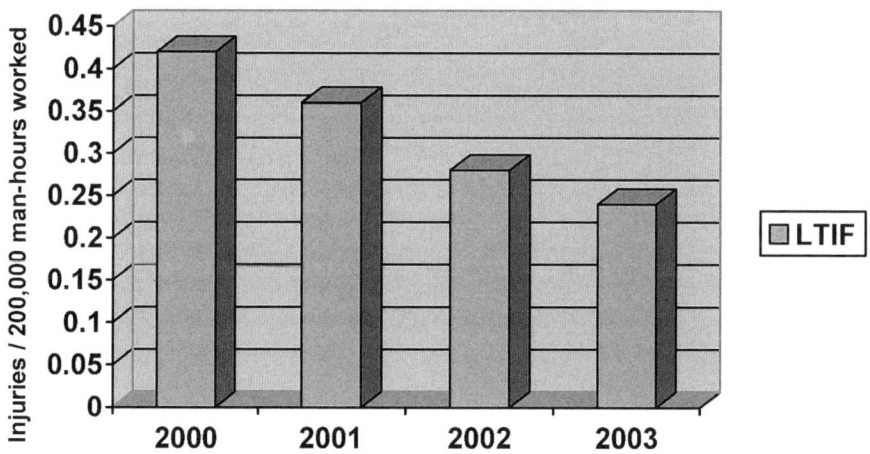

FIGURE 12. Industry sector LTIF – January 2000 to December 2003.

to be expressed in the same units as those used by Blue Ocean Off-shore and Green Sea Offshore.

From Figures 10, 11 and 12 it is evident that:

- the overall trend of LTIs in the Offshore sector of the shipping industry is one of decreasing frequency; and
- the LTI frequency of both case-study companies is approximately half of that for the industry sector in general.

From this quantitative examination of both companies' accident statistics no significant cultural inferences could be drawn. However, it is significant that since the introduction of the ISM Code and the STCW Convention the sea-going and shore based personnel in both case study companies have undergone similar training and safety indoctrination, which implies that the training provided may well be the reason for the good and improving safety statistics in both companies, thus lending support to the argument for more emphasis on education and training in preference to stricter enforcement of safety regulations in order to improve maritime safety standards.

TRAINING AND EMPLOYMENT POLICIES

There were differences between the training philosophies of Blue Ocean Offshore and Green Sea Offshore. In Blue Ocean Offshore training was provided on an *ad hoc* basis whereas in Green Sea Offshore training was provided on a planned basis monitored by a training officer. The main reason for the difference was the different employment philosophies of the two companies.

Blue Ocean Offshore

Crewing agents were the principal source of ships' officers and ratings in Blue Ocean Offshore, although a small number of senior officers were directly employed by the company. But in either case, the personnel were employed on a single voyage contract basis. No sea-going personnel were employed using long-term or 'rolling' contracts of employment. As a consequence of this employment philosophy, sea-going personnel were free to move on to another employer after completing a single voyage with the company.

Senior management had put forward a proposal to employ a human resources consultant for a period of two years acting at a senior level to develop a personnel management system that would encourage the most capable and promising sea-going staff to stay with the company. That would have ensured that finite resources spent on training would not be expended on people who did only one or two voyages with the company and then moved on to another company.

The HSE manager supported the proposal noting that employing personnel on a casual basis to crew company ships made continual improvement of safety standards particularly difficult. However, the commercial manager resisted the proposal, seeing the employment of a human resources consultant as unnecessary and a potential drain on corporate resources.

The proposal was ultimately not advanced any further, resulting in a continuing lack of direction in the HR department and reliance on crewing agencies to provide personnel who frequently had undergone only basic safety training. Company training was not methodical and was provided only on an *ad hoc* basis as illustrated by the following examples.

Two new vessels fitted with Azimuth thrusters rather than

170

conventional propellers and rudders were to be delivered to Blue Ocean Offshore. Operations staff realised that Masters being appointed to the vessels would require training in the requisite ship-handling techniques, which are quite different from those used on conventional vessels. A five-day course for three Masters and two Operations Superintendents was arranged with the Singapore Port Authority and the vessels went into service without mishap as far as ship handling was concerned.

Similarly, both new vessels had Wartsila main engines, and because no other vessels in the fleet were equipped with Wartsila engines Blue Ocean Offshore sent three selected Chief Engineers on familiarisation courses at the manufacturer's factory. This ensured that the vessels went into service without mishap as far as mechanical propulsion was concerned.

In contrast to the foregoing, each of the two new vessels was equipped with a fast rescue craft (FRC) the handling of which also requires specialist training. None of the crew appointed to the first vessel to be delivered had the requisite training and when the FRC was launched for testing during vessel acceptance trials, it was only by good fortune that the ship's Bo'sun in charge of the operation narrowly escaped severe physical injury. The subsequent hazardous incident report was sufficient to identify the need for FRC training and this was subsequently provided.

Green Sea Offshore

Green Sea Offshore operated in the British sector of the North Sea from UK ports and its vessels were British registered. Therefore regulations imposed by the British Maritime Administration and by seafarers' trade unions provided a strong incentive for the company to employ British seafarers in preference to foreign nationals.

The company addressed this constraint by using its own wholly owned employment agency to oversee and administer not only the supply of seafarers to Green Sea Offshore but also their training. The agency had a full-time training officer responsible for sending sea-farers on mandatory and elective training courses including the following:

- One week mandatory basic training course approved by the Marine and Coastguard Agency (MCA) for new entrants to the shipping industry covering Fire Fighting, First Aid, Sea Survival and Personal Safety & Social Responsibility.

- One week obligatory course approved by the Offshore Petroleum Industry Training Organization (OPITO) covering First Aid and Casualty Recovery by Fast Rescue Craft (FRC).
- Five-day Advanced Medical Course (AMA). To attend this course the candidate must first have passed a Green Sea Offshore multiple-choice test and have been recommended by the college.
- All deck crew were obliged by Green Sea Offshore to attend a two-day FRC course for the award of Boatman Certificate.
- Selected deck crew attended an additional five-day course for the award of FRC Coxswain Certificate.
- After at least three months' experience, selected FRC coxswains could attend a five-day Daughter Craft Coxswain Certificate course qualifying them to cox an enclosed rescue boat steering by compass and GPS only.

With the exception of the AMA course, the foregoing courses were refreshed by periodic training on board conducted by a specialist training company, while AMA refresher courses were carried out ashore. An ongoing training programme was also in place, validated by OPITO, with exercises covering six modules set by the specialist training company. Each module had questions for selected crew members and the responses together with the record sheet were returned to the training company for marking and identifying any areas of concern. All the relevant information was then entered into an agency database.

During a relational discussion, the company's training officer affirmed the effectiveness of the training by means of an illustration. Throughout the basic first aid course a deckhand complained for the entire week about having to attend the course. The training officer told him that once he had successfully completed the course it might one day be useful. About three months later the deckhand told the training officer that his newly gained knowledge had indeed proved useful. While the deckhand was on leave his father had suffered a heart attack, and using the skills he had learned on the course the deckhand had been able to keep his father alive until the paramedics arrived.

Cultural inferences

Of interest and important to note in the context of this study is that while both companies sourced their sea-going personnel through crewing agencies, sea-going personnel in Green Sea Offshore were effectively company employees because the company owned the agency. Sea-going personnel in Blue Ocean Offshore on the other hand were essentially contract labour hired on a per voyage basis.

Attempts made by Blue Ocean Offshore to engage Filipino officers on long-term contracts were abandoned without success after two years of negotiation. The Filipino officers preferred to be employed on short-term contracts and remain free to join another company at the end of each voyage. This was the case even with officers who had worked for the company for a number of years and were considered more as permanent employees than casual employees.

From relational conversations it is apparent that similar situations have been observed in other companies employing both British and Indian seafarers, such as Mobil Shipping Company Ltd. In that company British seafarers were content to sign long-term employment contracts but Indian seafarers preferred to be employed on a per voyage basis. This tends to show either an independence of thought on the part of Asian seafarers in general or an inherent distrust by Asian seafarers of shipping companies as employers.

COMPETENCIES, UNDERSTANDING AND REGULATION

From the ISM Code model (see Figure 9), three of the four constraints and pressures to be found at the level of operational safety are the cultural norms and heuristics of the operations staff, their competencies and understanding, and how they perceive the degree of regulation that currently exists with regard to implementing the ISM Code in accordance with the guidelines. Consequently, during Phase I of the empirical study, all twelve of the key informants interviewed were asked three definitive questions regarding operational safety relevant to those particular constraints and pressures. From analysis of the detailed questions and responses, the general consensus of opinion was as follows:

- Question One: From your experience, what do you believe is the most effective way of improving overall ship operating standards?
  Summary of responses:
  - More emphasis on professional training
  - More emphasis on safety training and quality control
  - Stricter enforcement of existing regulations.
- Question Two: How do you perceive the degree of regulation of the shipping industry?
  Summary of responses:
  - Ten of the twelve respondents thought the shipping industry was already sufficiently regulated
  - One respondent thought the industry was over regulated
  - One respondent thought the industry was under regulated.
- Question Three: What do you believe are the two most significant potential hazards to safety in the shipping industry?

  Summary of responses:
  Although couched in different terms, one theme was common to the responses of all interviewees: too low a standard of education and training.

From the foregoing it is apparent that the respondents across both the cultural and managerial divides in each company were of the opinion that the shipping industry is already sufficiently regulated and what is required to improve overall safety standards in the industry is greater emphasis on education and training.

SUMMARY

In this chapter operational safety was examined, both from a quantitative aspect and a qualitative aspect. The chapter began with a review of each company's safety management system. Both had elected to use a closed cycle SMS as described in Chapter 4 and the SMS of both companies comprised a comprehensive documented system in a tiered format under three heads:

- Company Policies
- Company Procedures
- Company Forms.

Where the two companies' systems were completely different was in their approach to maintenance of the vessels. Although each company had a planned maintenance system, Blue Ocean Offshore operated a computer-based system that put the onus of all operational maintenance on ships' staff, whereas Green Sea Offshore had a documentary system that charged shore-based staff with the responsibility for all maintenance other than routine checks.

An examination of each company's accident reporting and follow-up procedures was followed by a comparison of their accident statistics, which were seen to be somewhat similar and much better than the average for the industry sector.

The training and employment policies of both companies were then reviewed and a connection between the two policies was established. The different approaches to employment and training by the two companies were highlighted and examples were used to illustrate the effects of the different policies.

Finally, attention was drawn to the fact that although the differences between the training policies employed by the two companies were quite distinct, the key interview responses reported in the previous chapter indicated that all interviewees were agreed that raising the standards of education and training was the key to improving standards of safety in the shipping industry world-wide.

12

Case Study Phase Three: Behavioural Safety

Steady as she Goes

THE APPROACH

Analysis of behavioural safety at level 5 of the safety hierarchy involved measuring salient psychological and cultural dimensions of the seafarers engaged to crew the vessels of the two case study companies and comparing those dimensions with each individual's educational background and experience.

Both quantitative and qualitative techniques were used to gather the information necessary to make the comparisons. The former involved distribution of a two-part questionnaire-style survey among representative samples of British and Filipino seafarers to measure their locus of control orientation, rank, experience, and professional and academic qualifications, while the latter involved visiting ships, observing their overall condition and holding relational conversations with crew members to gain an impression of how they viewed the introduction of the ISM Code and the ever-increasing emphasis on safety.

The results of these two modes of enquiry are examined in this chapter. First, the responses to the questionnaires are examined statistically to identify any significant differences or correlations between the two samples. Second, an illustrative account is given of the salient points noted during the relational conversations and observations carried out on board the ships.

THE QUESTIONNAIRE

The questionnaire comprised two parts. The first part contained several questions of fact used to establish each respondent's rank, experience, professional qualifications and academic qualifications.

The second part comprised a psychometric test designed to establish each respondent's locus of control orientation (LOC). This part of the questionnaire contained 29 pairs of statements based on Rotter's[1] inventory with the wording slightly modified to lend an overall maritime flavour to the questionnaire. Respondents were asked to select the one statement in each pair that best described their feelings. Of the 29 pairs of statements, 23 contained choices with an internal/external bias and the other six were filler statements, one of which was used to gauge each respondent's feelings about the increasingly high degree of safety regulation now in force in the shipping industry.

QUESTIONNAIRE RESPONSE ANALYSIS

The questionnaire responses were first tabulated using Microsoft Excel software and the datasets produced were then subjected to statistical analysis with the analytical platform provided by an integrated SAS computer package and the resulting dataset exported to an Excel file for presentation in the format used for the statistical tables below.

Because the population distribution of the samples could not be assumed to be normal and the data produced by the questionnaire responses involved predominantly category data and ranked data, the statistical analysis involved the use of non-parametric tests.

The tests fell into three categories and in each category non-parametric methods were used for computer modelling and subsequent analysis as follows:

- Tests of differences between groups (independent samples):
    - Mann-Whitney U Test for independent samples
    - Kruskal-Wallis analysis of ranks for multiple groups.
- Tests of differences between variables (dependent samples):
    - Wilcoxon Matched Pairs Test to compare two variables measured in the same sample

- – Chi-square test for variables of a dichotomous nature
- Tests for relationships between variables (categorical in nature):
  - – Chi-square test for the relationships between the variables
  - – Spearman Correlation Coefficients to express those relationships.

The analysis enabled a picture to be built up of how the samples of Filipino and British seafarers differed. First, correlations were examined, both overall and then for each nationality. Second, the two samples were compared to see if there were any significant differences between the nationalities using the measured variables. Third, analysis was carried out on the component parts of the questionnaire to detect any unusual differences between the groups.

Basic statistics and correlation matrices are presented below together with descriptive analysis of the results.

BASIC STATISTICS AND CORRELATIONS

Combined samples

Table 12 provides a picture of the measured variables of the combined samples overall, while Table 13 presents the results of analysis of the basic statistics contained in Table 12.

TABLE 12. Basic statistics (combined samples)

Variable	N	Mean	SD	Median	Minimum	Maximum
LOC index	60	34.06	17.37	30.43	4.35	82.61
Rank	60	2.75	1.27	3.00	1.00	5.00
Experience	60	4.52	1.17	5.00	2.00	6.00
Professional qualifications	60	4.12	2.18	3.00	1.00	9.00
Academic qualifications	60	3.85	0.88	4.00	1.00	5.00

TABLE 13. Statistical correlations (combined samples)

	LOC index	Rank	Experience	Professional qualifications	Academic qualifications
Spearman Correlation Coefficients, $N = 60$					
Prob > |r| under H0: Rho = 0					
LOC index	1	−0.06325	0.03799	−0.08815	−0.18662
p		0.6312	0.7732	0.503	0.1534
Rank	−0.06325	1	0.35927	0.92981	0.48028
p	0.6312		0.0048	<0.0001	0.0001
Experience	0.03799	0.35927	1	0.29615	−0.19434
p	0.7732	0.0048		0.0216	0.1368
Professional qualifications	−0.08815	0.92981	0.29615	1	0.49258
p	0.503	<0.0001	0.0216		<0.0001
Academic qualifications	−0.18662	0.48028	−0.19434	0.49258	1
p	0.1534	0.0001	0.1368	<0.0001	

From Table 13 it is evident that none of the variables is significantly correlated with the LOC Index. However, rank is significantly correlated with:

- experience ($p = 0.0048$);
- professional qualifications ($p < 0.0001$); and
- academic qualifications ($p = 0.0001$).

Logically, this is to be expected, as also is the fact that:

- experience is significantly correlated with professional qualifications ($p = 0.0216$); and
- academic qualifications are significantly correlated with professional qualifications ($p < 0.0001$).

Similarly, we should not be surprised to find that academic qualifications are not correlated with experience ($p = 0.1368$).

The fact that these results are coincident with expectations may serve to reassure the researcher of the reliability of the computerised statistics program!

British seafarers

Table 14 gives a picture of the measured variables of the samples of British seafarers and from Table 15 it is evident that, as with the combined sample, none of the variables in the sample of British seafarers is significantly correlated with the LOC index.

However, unlike the combined sample, rank and experience are

TABLE 14. Basic statistics (British sample)

Variable	N	Mean	SD	Median	Minimum	Maximum
LOC Index	30	39.28	19.32	41.31	4.35	82.61
Rank	30	3.03	1.38	3.00	1.00	5.00
Experience	30	4.90	1.09	5.00	3.00	6.00
Professional qualifications	30	4.83	2.35	6.00	2.00	9.00
Academic qualifications	30	3.60	1.10	4.00	1.00	5.00

TABLE 15. Statistical correlations (British sample)

	LOC index	Rank	Experience	Professional qualifications	Academic qualifications
Spearman Correlation Coefficients, $N = 30$ **Prob > |r| under H0: Rho = 0**					
LOC index P	1	−0.22687 0.228	−0.12186 0.5212	−0.26845 0.1515	−0.21509 0.2537
Rank P	−0.22687 0.228	1	0.06318 0.7401	0.92568 <0.0001	0.78709 <0.0001
Experience P	−0.12186 0.5212	0.06318 0.7401	1	−0.02066 0.9137	−0.25581 0.1724
Professional qualifications P	−0.26845 0.1515	0.92568 <0.0001	−0.02066 0.9137	1	0.76217 <0.0001
Academic qualifications P	−0.21509 0.2537	0.78709 <0.0001	−0.25581 0.1724	0.76217 <0.0001	1

also not significantly correlated although academic qualifications are significantly correlated with professional qualifications. This may be indicative of the practice in the British merchant marine of training school leavers as officer cadets rather than ratings working their way up from the lower ranks to become officers. In the British system, young people must have attained a stipulated minimum academic standard prior to entering the shipping industry as officer cadets. They then undergo training for further academic awards and professional certificates.

Filipino seafarers

Table 16 provides a picture of the measured variables of the samples of Filipino seafarers and in Table 17, as with the previous two samples, none of the variables is significantly correlated with the LOC Index.

However, unlike the sample of British seafarers, rank and experience are highly correlated ($p < .0001$) among the sample of Filipino seafarers. Also of note is that there is no significant link between the attainment of professional qualifications and academic qualifications ($p = .1812$). This may be a reflection of a culture of moving up the ranks. Contrary to the practice in the British merchant marine, it is quite usual for Filipino seafarers, especially those employed in the offshore sector of the shipping industry, to work their way up through the ranks from ratings to senior officers during their careers.

TABLE 16. Basic statistics (Filipino sample)

Variable	N	Mean	SD	Median	Minimum	Maximum
LOC index	30	28.84	13.57	30.43	4.35	60.87
Rank	30	2.47	1.11	2.50	1.00	5.00
Experience	30	4.13	1.14	4.00	2.00	6.00
Professional qualifications	30	3.40	1.75	3.00	1.00	8.00
Academic qualifications	30	4.10	0.48	4.00	3.00	5.00

TABLE 17. Statistical correlations (Filipino sample)

Spearman Correlation Coefficients, $N = 30$ Prob > |r| under H0: Rho = 0					
	LOC index	Rank	Experience	Professional qualifications	Academic qualifications
LOC index	1	−0.06163	0.05314	−0.07789	−0.11635
p		0.7463	0.7803	0.6824	0.5404
Rank	−0.06163	1	0.69438	0.87957	0.09921
p	0.7463		<0.0001	<0.0001	0.6019
Experience	0.05314	0.69438	1	0.57151	−0.03765
p	0.7803	<0.0001		0.001	0.8434
Professional qualifications	−0.07789	0.87957	0.57151	1	0.25087
p	0.6824	<0.0001	0.001		0.1812
Academic qualifications	−0.11635	0.09921	−0.03765	0.25087	1
p	0.5404	0.6019	0.8434	0.1812	

MANN-WHITNEY U TEST

For reasons listed earlier the Mann-Whitney U Test was used rather than its parametric counterpart in order to test the differences between the samples of British and Filipino seafarers. The results of the analysis are shown in Table 18.

TABLE 18. Significance of differences between samples

Observed Variable	British mean	Filipino mean	Mann-Whitney U Test
Rank	3.03	2.47	0.0804
Experience	4.90	4.13	0.0146
Professional qualifications	4.83	3.40	0.0284
Academic qualifications	3.60	4.10	0.0776
LOC index	39.28	28.84	0.0452

The most surprising fact that emerges from the comparison is that British seafarers have a significantly higher locus of control index than Filipino seafarers (39.28 vs 28.84) ($p = 0.0452$). The greater the LOC index the more external the LOC orientation and the lower the LOC index the more internal the LOC orientation. From the analysis, therefore, the locus of control of Filipino seafarers is significantly more internally oriented than that of British seafarers.

This finding is contrary to that anticipated from the literature review, specifically in Chapter 5 where it was argued that the general consensus of opinion was that a significant positive correlation exists between individualism and internal locus of control, from which it could be assumed that British seafarers would have a more internal LOC orientation than Filipino seafarers, the former belonging to a more individualistic society than the latter.

With regard to the other observed variables in Table 18, although Filipino seafarers have higher levels of academic qualifications than British seafarers the difference is not statistically significant ($p = 0.0776$). Similarly, although British seafarers have a slightly higher mean rank than the Filipino seafarers, the difference is not statistically significant ($p = 0.0804$). These differences may simply reflect the make-up of the two samples, the populations of which were randomly selected. This may also account for British seafarers having a higher level of professional qualifications than Filipino seafarers ($p = 0.0284$), higher ranks requiring higher levels of professional qualifications.

Although random selection of the sample populations may also have a bearing on the fact that the British seafarers are more experienced than Filipino seafarers ($p = 0.0146$), this difference may in fact reflect a demographic difference between the two groups, British seafarers in general having a high age profile.[2]

COMPONENT ANALYSIS

To detect any unusual differences between the two groups of seafarers at the level of the individual components that comprised the LOC questionnaire, the responses to the individual alternatives were tested. A chi-square analysis was made of the cross-tabulated frequencies of each selected alternative compared to the expected frequencies. The cumulative difference over all the cells in the cross tabulation was then used to identify whether there were any significant differences.

To compensate for the increased probability of making Type I errors due to the sheer number of repeated comparisons, a Bonferroni adjustment[3] was applied. In order to keep the overall error rate to an acceptable level of 5%, the α-level of 0.05 was divided by the number of measured responses (23) so that for any one comparison to be considered significant the obtained p-value would have to be less than 0.002174.

The only statistically significant difference found at the Bonferroni adjusted significance level was in the responses to alternative 22 where $p < 0.00001$. Here the choice of responses was between:

- with enough effort we can wipe out political corruption; and
- it is difficult for people to have much control over things politicians do in office.

The response frequencies are shown in Table 19, from which it is evident that there is a striking difference between the number of Filipinos who think they can work together to wipe out political corruption and the few British seafarers of like mind.

Of course, the political systems in the United Kingdom and the Philippines are quite different from each other as is the attitude towards direct action. This may be illustrated by quoting from a letter received by the Blue Ocean Offshore Crewing Supervisor from

TABLE 19. Responses to Alternative Choice Question 22

a22(a22)	Nationality		Total
	British	Filipino	
With enough effort we can wipe out political corruption	9	27	36
It is difficult for people to have much control over the things politicians do in office	21	3	24
Total	30	30	60

the Second Engineer of one of the company's smaller vessels, referred to here as *Blue Ocean IV*. The letter read as follows:

Sir,

Permission to go home as soon as Blue Ocean IV arrives Singapore. Big problem encountered in my home. Part of my residential house damaged – somebody throw PC Grenade. Ended – caught the suspect behind political rivals – and put in jail. Promise to return if company need me to continue my second contract. Forgive me for 6 days leave. Willing to back soon to Blue Ocean IV.

With respect,
2/E Signature
Blue Ocean IV

Apparently, the Second Engineer's wife was a local politician and someone who disagreed with her political views took direct action to make his or her feelings known by throwing a hand grenade into the house, demolishing a part of the building but fortunately without injuring anyone.

Also of interest was the perfect correlation between the samples in the responses to alternative 15, as shown in Table 20. This indicates that there was absolutely no difference between the sample groups in the belief that luck had nothing to do with getting what they want.

This may be an area where, due to their occupation, seafarers hold common beliefs and values regardless of their national cultures.

TABLE 20. Responses to Alternative Choice Question 15

a15(a15)	Nationality		Total
	British	**Filipino**	
In my case, getting what I want has little or nothing to do with luck	26	26	52
Many times we might just as well decide what to do by flipping a coin	4	4	8
Total	30	30	60

CHRONOLOGICAL CORRELATION

In Chapter 5 it was deliberated whether or not it might be possible by measuring the locus of control of individuals and comparing it with their educational attainments, cultural backgrounds and work roles to discern whether or not locus of control orientation has situational and chronological aspects, moving from external to internal as people gain knowledge and experience or gain promotion, i.e. as they gain greater control over their lives.

From the foregoing analyses when the correlation of rank and experience in each of the two samples is compared with the respective LOC index it does at first appear that such a connection might be made. Consider the pertinent data:

- For British seafarers rank and experience are not significantly correlated: ($p = 0.7401$) and the mean LOC index is 39.28.
- For Filipino seafarers rank and experience are highly correlated: ($p < 0.0001$) and the mean LOC index is 28.84.
- The LOC index scale is such that the higher the index score the more external the orientation and the lower the index score the more internal the orientation.

It is tempting to infer, therefore, that the greater the correlation of rank and experience the lower will be the LOC index and correspondingly more internal the locus of control orientation, thus confirming the hypothesis that locus of control has situational and chronological aspects. Intuitively it seems logical that as people progress through life and gain increased knowledge, more skills and greater control over their lives, their locus of control orientation would tend to move from external to internal.

However, this is an area of some uncertainty. While there is a statistically significant difference between the LOC scores of Filipino and British seafarers, this difference cannot be definitely attributed to rank or experience. There is a possibility that it is, but there are insufficient data to support that theory.

Such uncertainty is generally a feature of small datasets due to the fact that a few unusual observations have a much larger than warranted influence on a statistical test. With a large dataset the background noise loses its influence. A much larger sample normally removes all these inconsistencies. However, even then correlation should not be mistaken for causation.

Without completing a large empirical study it would not be possible to say that a higher rank or greater level of experience would cause a lower LOC. Even then, if a significant correlation was measured, some additional testing in the field would need to be carried out to establish a positive link determining cause and effect. This would be a fruitful field for further research.

Meanwhile, the most positive assumption that can be made from the present statistical analysis is that it provides tenuous evidence of a link between LOC and experience, thus providing some weight to the argument for greater emphasis on education and training as a means of developing internal motivational cognition in operational situations, particularly emergencies.

RELATIONAL DISCUSSIONS AND OBSERVATION

Statistical analysis of the responses to the questionnaire-style survey of two samples of seafarers, one from each cultural group, was one of two methods used in the third stage of the empirical research. The other method was qualitative, comprising documentary review, relational discussions and observation.

Although the seafarers employed by the two case study companies were from two completely different cultural backgrounds, the empirical research revealed a surprising similarity in their attitudes to various operational matters as illustrated in the following vignettes.

Customer satisfaction

Both the Filipino and the British seafarers felt it was their duty to ensure that the end user of their vessels' services, whether time charterer, voyage charterer or cargo receiver, was satisfied with the performance of the vessel and its crew. Sometimes the desire to keep the customer happy was so strong that safety became a secondary consideration. Two examples will help to illustrate this point.

The first example concerns a Blue Ocean Offshore supply vessel with a Filipino Master and Chief Mate. The remainder of the crew comprised Filipinos and Indonesians. The vessel loaded a cargo of diesel oil, fresh water and pipes in a port on the east coast of Malaysia to take to an oil rig 150 miles offshore. The voyage from the port to the oil rig would take approximately 14 hours.

Just before the vessel reached the rig, however, instructions were received to return to port, offload the cargo and proceed directly to Singapore for dry-docking and survey. The Master turned the vessel around and started to return to port. The weather was moderate to rough and seas were breaking over the vessel's decks. Two hours later the vessel lost all power and was floundering with no motive power at night, in a shipping lane for several hours before the Chief Engineer managed to restart the generators and engines.

Subsequent investigation showed that having first loaded a part cargo of diesel oil and fresh water in designated tanks, the vessel then began taking on board a deck cargo of pipes. Charterers wanted to load as many pipes as possible for delivery to the oil rig and in order to please the charterers the Master took on board so many pipes that his vessel was in fact overloaded, considerably reducing the freeboard. As a result, when the vessel left port and encountered choppy seas, water entered the vessel's bunker fuel tanks through vents on deck and eventually the water in the fuel caused the generators and main engine to splutter to a halt.

In reviewing the case with the Master it became apparent that he believed he had acted reasonably in allowing his vessel to be overloaded because in his opinion it was of the utmost importance both to avoid confrontation and to please the clients in their dual role of charterer and customer.

The second example concerns a number of potential collisions by Green Sea Offshore vessels with oil rigs and production platforms in the North Sea. Reports of such hazardous occurrences were not infrequently submitted by British Masters of North Sea vessels and subsequent investigation found that in many cases the root cause of the incidents was the Master's desire to please the charterer by loading or offloading cargo at the offshore installations in circumstances such as bad weather when the Master should in fact have refused to work cargo at all.

It could be argued that the first incident was an example of two of Hofstede's[4] cultural dimensions, power-distance and uncertainty avoidance, influencing the Master's judgement. However, that would not account for the reported hazardous occurrences in the North Sea. It is more likely that in both situations the Masters were either overconfident in their abilities or lacked sufficient knowledge or experience to recognise the hazards confronting them.

Exercising a Master's over-riding authority

The ISM Code makes it very clear, as does the SMS of each case study company, that in circumstances where a decision has to be made concerning the safety of a vessel or her crew the Master has over-riding authority. But the exercise of authority is invariably accompanied by responsibility and exercising over-riding authority implies taking ultimate responsibility for one's decisions.

Therefore, while cultural factors may to some extent influence decisions made concerning the safety of a vessel and her crew, it is more likely that pragmatic considerations and public reaction will play a greater part in the decision-making process, as illustrated by the following further two examples.

The first concerns a production platform located offshore Vietnam, a developing country with a centrally controlled economy where public pressure to raise safety standards is not significant. The physical location of the particular production platform and the position of its crane used for transferring cargoes to and from supply boats were such that the supply boats had to work against the current and, dependent upon the time of day, sometimes also against the tide. But despite several hazardous occurrences being reported and two actual collisions with the platform being recorded, the supply boats continued to service the platform.

Several Filipino Masters had discussed with various Offshore Installation Managers (OIMs), all Europeans, the dangers posed by the current, the tide and the crane location but were told that there was nothing that could be done. Reluctance on the part of the Filipino Masters to argue the case was understandable not only because of their cultural preference to avoid confrontation but also because of a perceived lack of support from shore-based management coupled with public indifference. The Masters knew that such operations had previously been carried out without mishap and realised that refusal on their part to service the platform would most probably lead to cancellation of the vessel's charter and another company taking it over with subsequent loss of their jobs. They knew also that in the event of an accident there would be no public outcry in a country with a centrally controlled economy and government controlled media.

The second example concerns a supply boat operation to a production platform in the North Sea, and serves to illustrate how in northern Europe public perception of the hazards of oil and gas production, heightened by the 1988 *Piper Alpha* disaster that resulted in the deaths of 167 men, requires operators of offshore installations to take into consideration public reaction should an accident occur due to unacceptable risk taking.

During a relational discussion the British Master of a 3860 horsepower supply boat was asked whether he felt his vessel was sufficiently powered for North Sea operations, especially if he were to find himself in a difficult situation. His response was that his vessel was sufficiently powered because one should not allow oneself to get into a difficult situation in the first place. By way of support for his argument he pointed out that a company now operating many of the former British Petroleum marginal platforms in the North Sea had agreed that as a measure to reduce the risk of accidents supply boats would not be required to do up-tide or weather-side operations.

Most OIMs and many other staff on offshore installations have worked in various different areas of the world and ultimately the safety criteria are similar wherever the offshore installations are located. It is arguable, therefore, that the difference between the attitudes of the Masters of the supply boats in the South China Sea and those in the North Sea, and also the difference between the attitudes of the OIMs on platforms offshore Vietnam and those on platforms offshore UK, were due not to the influence of their respective national cultures but to the different styles of prevailing regional governments, the influence of the local and regional media and consequently the effectiveness of public pressure.

Condition of vessels

Several vessels of each case study company were visited on a completely random basis dependent upon their availability. Some were visited while undergoing repairs in shipyards and others were visited during operational port turnarounds.

The majority of vessels operated by both companies and all of the vessels visited were between 20 and 30 years of age, which is quite old in shipping terms. However, from observation it was apparent that the vessels visited were all well maintained. This was substantiated by inspection of documentary records which showed that all the vessels were in class with reputable classification societies and had current trading certificates including safety management certificates. From observation and relational conversations it was also established that all of the vessels had at least the required minimum safety equipment and many of the vessels actually carried safety equipment over and above the required minimum.

The standard of operational maintenance and overall appearance of the vessels depended upon a number of considerations: e.g. whether the vessel was in a shipyard or operational at the time of the inspection; whether it had been operating on safety stand-by or carrying out anchor handling duties; whether the vessel had been operating in heavy seas which precluded painting and similar maintenance work or whether it had been sailing in calm water. Overall however, the vessels were all found in an acceptable condition and there was no significant difference between the observed standards of operational maintenance or overall appearance of the vessels of either company. The different economic circumstances of the case study companies were not reflected in the standard of vessel maintenance.

Response to increased emphasis on safety

In relational discussions with sea-going staff most expressed the view that improving safety at sea was a laudable objective and this feeling was substantiated by a response to one of the alternate choice statements contained in the LOC questionnaire distributed to both sample groups of seafarers.

The questionnaire contained six pairs of alternate choice statements that were not relevant to the main objective of measuring each respondent's LOC orientation. However, one of them was relevant to the general perception of seafarers to the increasing emphasis on safety in shipping operations. The respondents were asked to choose between the following statements:

a. There is too much emphasis on safety on board ships nowadays.
b. Emphasising safety on board ship helps to develop a team spirit.

Of the 60 respondents a total of 4 declined to respond, only 8 selected option (a) and 48, i.e. 80%, selected option (b). Such a high percentage is considered to be statistically significant.

The general attitude of Masters and Chief Engineers expressed during relational discussions was summed up by the Master of a North Sea supply boat when he said that the introduction of the ISM Code was a good thing in so far as it requires not only procedures but also authorities to be documented, that it has undoubtedly helped to improve safety standards, but the system needs to be refined to reduce paperwork.

SUMMARY

The research format described in this chapter was both quantitative and qualitative in nature and focused on the seafarers employed by the two case study companies.

Findings of quantitative research

Statistical analysis of the responses to a questionnaire distributed among two representative samples of the sea-going personnel employed in each company found that:

- LOC index was not significantly correlated to any of the variables.
- Rank and experience were not significantly correlated in the case of British seafarers but were highly correlated in the case of Filipino seafarers. This may be a reflection of the difference between the traditional officer-training methods used in the two sample groups.
- Filipino seafarers had a lower LOC index (i.e. greater internal LOC orientation) than the British seafarers, who had a greater external LOC orientation. This finding together with the correlations of rank and experience in each sample group may indicate that locus of control has chronological aspects, an individuals' locus of control tending to move from external to internal as they progress through life gaining increased knowledge, more skills and greater control over their lives. However, there are

insufficient data to confirm the theory and more research would need to be done in this area to establish a positive link.

- A chi-square analysis of the cross-tabulation frequencies of each pair of selected alternative responses to the questionnaire compared to the expected frequencies detected only two unusual results:
  - First, the Filipino seafarers were more certain than the British that they could work together to wipe out political corruption, which may reflect the difference between their LOC indices or simply the different political systems in their home countries.
  - Second, both groups were strongly convinced that luck had little or nothing to do with getting what they want. This may indicate that due to their occupation seafarers hold some common beliefs and values regardless of their national cultures.

Findings of qualitative research

Based upon information drawn from documentary review and key informant interviews supplemented by relational conversations and observation, some vignettes were used to illustrate that what on the one hand could be interpreted as the influence of cultural dimensions could also be interpreted on the other hand as the result of a lack of knowledge or experience.

Vignettes were also used to illustrate that a Master's willingness to exercise his over-riding authority in cases where the safety of a vessel or its crew might be compromised could be interpreted on the one hand as being culturally influenced or on the other hand as being pragmatically driven.

Vessels of both companies were visited and seen to be well maintained and well equipped. The vessels were all in class and showed no signs of suffering from a lack of maintenance or lack of safety equipment as a result of economic restrictions.

During relational discussions ships' crews generally expressed their approval and acceptance of the increasing emphasis placed on improving safety standards at sea and this was supported by the 'filler' alternative choice statement in the LOC questionnaire.

13

Case Study: Comparative Safety Climate Review and Summary

End of Passage

This chapter comprises a synopsis of the findings from each area in which research was undertaken, first into the structure of the two case study companies, followed by safety management at levels 3, 4 and 5 of the safety hierarchy. The information from each area of research is then used to construct a table showing the strengths and weakness of the safety culture components of each case study company and hence provide a comparative measure of their safety climates.

THE COMPANIES

Chapter 10 compared and contrasted two companies that had similar corporate structures but operated in culturally different contexts and with different financial pressures. Blue Ocean Offshore was a financially heavily leveraged company that operated offshore support vessels in the Middle East and South East Asia, while Green Sea Offshore was a cash-rich company that operated offshore support vessels in the North Sea.

The corporate safety culture of each company had developed differently due to their different corporate histories. Blue Ocean Offshore had consciously developed the strong points of the safety cultures of companies it had taken over whereas Green Sea Offshore retained a safety culture that reflected the values of the company prior to it being taken over and subsequently expanded.

ORGANISATIONAL SAFETY MANAGEMENT

Chapter 10 also examined in succession each of the constraints and pressures identified as operating at safety level 3 of the ISM Code model (see Figure 9).

Research at this phase of the case study determined that while constraints and pressures resulting from legal and moral obligations, economic considerations and organisational and cultural norms impact upon organisational safety management, they can be suitably dealt with by good strategic and organisational management using standard management techniques, which in a culturally homogeneous company may be of an indigenous nature but in a culturally heterogeneous company require in addition well-developed cross-cultural management skills.

The possibility of correlation between an individual's locus of control and job function was identified as an area where future management research might usefully be undertaken.

OPERATIONAL SAFETY MANAGEMENT

Chapter 11 examined operational safety, first from a quantitative aspect by examining the accident statistics of both case study companies, and second from a qualitative aspect by examining each company's:

- documented systems of management; and
- salient human aspects, particularly employees' perceptions regarding safety and company training and employment policies.

Comparison of each company's accident statistics showed them to be somewhat similar to each other and much better than the average for the industry sector.

Both companies had elected to develop and implement a closed-cycle SMS and the two systems were similar in format and content. Where the systems differed was in their approach to maintenance of the vessels. Although each company had a planned maintenance system, Blue Ocean Offshore operated a computer-based system that put the onus of all operational maintenance on sea-going staff whereas Green Sea Offshore had a documentary system that charged shore-based staff with the responsibility for all maintenance other than routine checks.

Finally, a review of both companies' training and employment policies established that the former was largely dependent upon the latter. While the differences between the policies of the two companies were quite distinct, analysis of the key interview responses indicated that the importance of those policies was not lost upon senior management. The analysis also indicated that all interviewees were agreed that raising the standards of education and training was the key to improving standards of safety in the shipping industry world-wide.

BEHAVIOURAL SAFETY

The research format described in Chapter 12 was both quantitative and qualitative in nature and aimed at the seafarers employed by the two case study companies. The findings of the quantitative and qualitative research are comprehensively summarised on pages 192 and 193 so need not be repeated here.

QUALITATIVE ASSESSMENT OF SAFETY CLIMATE

As noted in Chapter 4, the most common method of evaluating the effectiveness of an organisation's SMS is recording accidents, lost time incidents and hazardous occurrences.[1] However, due to the quantitative nature of statistics they do not readily reflect either the maturity of a company's safety culture or the effects of human influences on the application of its SMS. Other performance indicators must be selected to determine such qualitative issues.

It was further noted in Chapter 4 that safety climate is best measured by perceptual audit,[2] and in Chapter 9 the following factors were identified as being fundamental in the establishment of an organisational safety culture:

- Senior management commitment
- Line management commitment
- Commercial management involvement

- Operational pressures versus safety
- Ship/shore safety communications
- Provision of safety resources
- Employment philosophy
- Training programmes
- Mutual trust between sea-going and shore staff
- Shared perceptions about safety.

These essential factors, most of which are highly value laden and hence potentially subject to cultural influences, were therefore examined during the empirical study and the research findings can be used to determine the relative maturity of a company's safety climate. The research findings relating to each factor are briefly summarised below.

Senior management commitment

From the key informant interviews, relational conversations, observation and document reviews, it was evident that the main strength of each company's safety profile was a sincere commitment by senior management to ensuring safe operation of its vessels. This finding was substantiated during relational conversations with the chairmen of the boards of directors of the conglomerates that owned Blue Ocean Offshore[3] and Green Sea Offshore.[4]

Line management commitment

While line management in both companies acknowledged the benefits of good safety management, their enthusiasm was tempered by the realities of implementing safety policies and ensuring that safe operating procedures were not only documented but also followed.

From a review of Blue Ocean Offshore's accident statistics and analysis of the key informant interview responses it was evident that a new SMS had been effective in reducing accidents and that near-miss reporting was being carried out by sea staff.

One discordant note was an initial perception, later reinforced from documentary review, that Blue Ocean Offshore's Operations Manager, although saying that he fully supported the company's safety policies, was in reality less committed to improving the overall safety performance of the fleet by implementation of an effective SMS, preferring instead to deal with problems as they arose rather

197

than taking proactive measures to prevent the problems from arising. He recognised perhaps that to develop and implement preventative measures to decrease the possibility of hazards eventuating or to minimise the effects of their occurrence requires a substantial amount of hard and disciplined work in the first instance.

It was also clear from key interview analysis and relational conversations with a wide spectrum of Blue Ocean Offshore staff that the vast majority of employees felt that the industry was already sufficiently regulated and that while there should be strict and more uniform enforcement of existing regulations the first priority was to place greater emphasis on both professional education and safety training.

Line management in Green Sea Offshore appeared to be fully supportive of the company's attempts to develop a safety culture within the organisation. It was evident, however, that line managers tended to identify with the corporate ethos that existed prior to the company being taken over rather than with the corporate identity of the new company.

However, training in internal auditing and ISM Code familiarisation had resulted in line management having a common understanding of the need not only to develop and implement an effective SMS but also to ensure its acceptance by sea-going staff and ensure continued improvement of the system.

Commercial management commitment

In the commercial departments of both case study companies the lack of safety training and only a passing acquaintance with operational risks was identified as a weakness in developing an overall safety culture.

None of Blue Sea Offshore's commercial personnel had a seafaring background or specific training in marine chartering. This occasionally led to substantial expenditure being necessary to retrieve a situation that chartering managers had not foreseen. Absence of safety training and lack of operational awareness also led to a policy differential between the company's marine personnel department and the desire of senior operations personnel to improve the overall operational and safety standards of ships' crews.

During relational conversations two points of view were expressed that indicated an overall weakness in Green Sea Offshore's commercial management commitment to safety. The first was that although overall management commitment to safety was good, there was a disconnect at local level in so far as the general manager did not have a marine background and had greater concern for commercial matters than for safety matters. The second was that while interdepartmental and ship–shore safety communications were generally very good, the commercial department were 'always ready to sell more than they have got' thus bringing additional pressures to bear on the operations staff.

Operational safety

The difference in employment philosophies and consequently training programmes of the two companies had little effect on the accident statistics in the short term but might ultimately be expected to favour Green Sea Offshore in the longer term.

The computerised planned maintenance system introduced by Blue Ocean Offshore initially met a lot of resistance from middle management and supervisory staff and there were many hurdles to be overcome before the system could be fully implemented. By dint of perseverance and the application of suitable education and training programmes, however, the system was eventually implemented and its effectiveness was reflected in a significant reduction in the number of days that vessels in the fleet were out of service for repair.

The document-based planned maintenance system used by Green Sea Offshore was very basic and straightforward but put additional responsibility on shore staff for vessel maintenance. The system assumed little competence on the part of the sea-going personnel and required them to operate rather than maintain the vessels. All maintenance other than that of a simple nature was scheduled and undertaken by shore-based staff. The reason for this approach may have been due to the origins of the fleet, the background of which was the fishing industry.

One area of interest noted during relational conversations with some Green Sea Offshore personnel, both sea-going and shore-based, concerned flag state inspections. The company's SMS was audited by the British Marine and Coastguard Agency (MCA) and one particular inspector whose cultural background was Pakistani was reportedly very difficult to deal with. It was felt by the Green Sea

Offshore personnel, all of whom had British cultural backgrounds, that because of his cultural heritage the MCA inspector:

- lacked both discretion and flexibility in dealing with non-conformities;
- was unable to make a decision that might later be shown to be wrong.

This is a classic example of uncertainty avoidance displayed by a person from a cultural environment with a high UAI (70 on Hofstede's scale) dealing with people from a cultural environment with a low UAI (35 on Hofstede's scale).

Safety communications

Both companies operated similar systems of reporting accidents and disseminating the information. There was a strong desire by senior management in both companies to ensure that accident reports were submitted as soon as possible after the occurrence of an incident. This was partly driven by a desire to receive the information from the vessel before hearing it from the client.

During the documentary review, inspection of reports from vessels gave no indication of any reluctance on the part of either British or Filipino seafarers to report either accidents or hazardous occurrences. This supported the assurances given by shore-based management and supervisory staff during the key informant interviews.

Safety resources

In both Blue Ocean Offshore and Green Sea Offshore the commitment of senior managers to safety was reflected in the provision of resources. The budgets for safety items were both generous and extensive, covering all areas of the safety spectrum, including:

- provision of safety equipment on board vessels;
- personal protective equipment for all personnel;
- safety training;
- monitoring and analysing accidents and near misses;
- upgrading vessels to meet new industry regulations and standards.

Nowhere was any evidence found in either company of economic considerations impacting upon the provision of safety resources.

Employment policy

While both companies sourced seagoing personnel through crewing agencies, those employed by Green Sea Offshore were virtually company employees because the company owned the crewing agency. The sea-going personnel of Blue Ocean Offshore on the other hand were effectively contract labour hired on a per voyage basis.

Attempts made by Blue Ocean Offshore to engage Filipino officers on long-term contracts were unsuccessful. Ostensibly the Filipino officers preferred to work on short-term per voyage contracts so that they were free to join another company at the end of each voyage. This may have been due to cultural factors, it may have demonstrated an independence of thought on the part of the Filipino seafarers or it may have been the result of inherent distrust of shipping companies as employers.

Training programmes

The difference in employment policies between the two companies was reflected in their training programmes. While Green Sea Offshore had in place a structured training programme under the direction of a training officer, Blue Ocean Offshore carried out training on an *ad hoc* basis initiated by perceived needs. Although Blue Ocean Offshore had a generous training budget, the company was understandably reluctant to provide more than the basic minimum training for staff who were reluctant to give long-term commitment to the company.

Mutual trust

Generally speaking there was a high standard of mutual trust between the seafarers and shore-based staff in both companies, evidenced by the forthright reporting of hazardous incidents.

But that mutual trust was qualified in some areas, particularly with regard to machinery and vessel maintenance. In both companies the shore-based staff felt that the seafarers were not sufficiently knowledgeable or experienced to carry out more than basic maintenance.

This was reflected in responses received during the key informant interviews when respondents were asked what they believed were the two most significant potential hazards to safety in the shipping industry. Although couched in different terms, one theme was common to the responses of all interviewees across both the cultural and managerial divides: too low a standard of education and training.

Shared perceptions of safety

Personnel who had attended training courses and seminars on the introduction and working of the ISM Code had very similar views concerning safety. Seafarers also had mandatory safety training and were therefore well versed in the objectives of the modern safety environment.

Accidents still happen however, and a review of the incident reports in both case study companies determined that many of the incidents were due not to a lack of safety training but to a lack of either forethought or knowledge. This indicates the need for more extensive vocational and professional training to establish heightened safety awareness and correspondingly improved perceptions of safety.

RELATIVE SAFETY CLIMATE

The foregoing summary of the findings of the empirical study carried out in the two case study companies provides a synopsis of what may be termed a perceptual audit of the two companies. As a means of measuring the safety climate of each company, and of each company relative to the other, the results of the perceptual audit have been used to construct a table indicating the strengths and weaknesses of each company in the various safety related areas of research. The tabulated results of the analysis provide:

- an indication of where each company needs to concentrate efforts to raise levels of organisational safety;

- a comparison of the maturity of the safety culture element in each company relative to the other.

A perceptual audit is a somewhat subjective means of measurement and hence open to cultural influences. Therefore, to be as objective as possible in establishing the safety climate of both case study companies, each of the essential factors was examined in turn with evidence drawn from the four research techniques utilised in the study (interview, documentary review, relational discussion and observation), noting any culturally influenced risk factors that prevailed and whether they exercised a positive or negative influence on safety organisation and performance.

A strengths and weaknesses analysis was then carried out, each essential factor being reviewed and given a low, medium or high rating in accordance with the perceived efficacy of organisational safety in that area of research, as shown in the safety climate comparator, Table 21.

Where an essential aspect has been given a high rating, then the constraints and pressures identified in the ISM Code model as impacting upon that particular element are considered to have been suitably dealt with by standard management techniques, combined with suitable education and training and the application where necessary of good cross-cultural management skills. Conversely, where an essential aspect has been given a low rating then the applicable constraints and pressures identified in the ISM Code model have had a notable influence and are not considered to have yet been suitably dealt with.

A similar technique is advocated by the Oil Companies International Marine Forum (OCIMF).[5] To achieve the main objective of safe management of oil tankers, the OCIMF identifies 12 essential elements of management practice each of which has a specific aim. Within each element key performance indicators (KPIs) are established and measured to determine whether or not the aim has been achieved, and hence how close the company is to achieving the overall main objective.

As discussed in Chapters 4 and 6, safety is a relative concept and the safety climate comparator illustrated in Table 21 could therefore be used to monitor improvements in a company's own safety climate by comparing periodic perceptual audits against each other. Alternatively, by establishing a benchmark standard against which

TABLE 21. Safety climate comparator

	Blue Ocean Offshore			Green Sea Offshore		
	Rating			Rating		
Essential factor	Low	Med	High	Low	Med	High
Senior management commitment			✓			✓
Line management commitment			✓			✓
Commercial management involvement	✓				✓	
Operational safety		✓			✓	
Ship/shore safety communications			✓			✓
Provision of safety resources			✓			✓
Employment philosophy		✓				✓
Training programmes	✓				✓	
Mutual trust between sea-going and shore staff			✓		✓	
Shared perceptions about safety			✓		✓	

to measure a company's safety climate, the comparator could be used as a relative safety culture maturity model (RSCMM) to measure the safety climate of other shipping companies relative to the benchmark standard.

Unlike Fleming's draft safety culture maturity model examined in Chapter 4, the model illustrated in Table 21 does not rely on organisations progressing through iterative stages, moving from one stage to the next only when the strengths and weakness at each stage have been built upon or removed respectively. Instead, the model uses a perceptual audit to identify any areas of weakness in safety management relative to a benchmark standard so that appropriate remedial action can be taken.

PART III

CONCLUSION

14

Review and Discussion

Voyage Report

It is hoped that this book has taken the reader on a voyage of discovery that has been both enjoyable and illuminating. Having now reached our destination we shall review in this final chapter the aims and objectives established at the start of our journey, the means by which those aims and objectives were addressed and the extent to which they were achieved.

THE AIMS AND OBJECTIVES OF THE BOOK

The aims and objectives of this book were:

1. to identify the human factors that impact upon the implementation of safety regulations in the world-wide shipping industry;
2. to determine what obstacles such impact may present to the development of safe practices and attitudes throughout the industry;
3. to carry out an empirical study to ascertain how companies actually address the obstacles to operational safety presented by the human factors and the degree of success they achieve;
4. to determine whether stricter enforcement of existing regulations or more emphasis on education and training is the better path to follow in order to raise safety standards throughout the world-wide shipping industry.

The first two objectives were achieved in Part I of the book by reviewing the relevant literature and using the results of the review to develop a model of the working of the ISM Code upon which were superimposed the various constraints and pressures emanating from human factors that act upon the development and implementation of safety regulations at various levels throughout the shipping industry, from governmental level to shipboard level.

The model presented a useful means of referencing the nature of the constraints and pressures occasioned by human factors such as cultural values and psychological dispositions, the organisational level of safety management at which those constraints and pressures are to be found, and the point at which an intervention needs to be aimed in order to address the negative effects of constraints and pressures operating at specific hierarchal safety levels.

The third objective was dealt with in Part II of the book. An empirical study using a comparative case study methodology was undertaken in two structurally similar but financially and culturally different shipping companies in order to acquire evidence of the actual impact of the constraints and pressures caused by human factors in those companies, the organisational methods used to deal with that impact and to what degree those methods were successful.

The case study also served to satisfy the fourth objective by testing the following alternative hypotheses:

> *Hypothesis A:* Cultural norms influence individual perceptions of safety. However, culture is learned and is not a static dimension. Therefore, by means of suitable education and training a common standard of safety can be achieved across a spectrum of individuals having diverse cultural norms.
>
> *Hypothesis B:* Cultural norms influence individual perceptions of safety. Therefore, rigorous enforcement of agreed rules and regulations is necessary to achieve a common standard of safety across a spectrum of individuals having diverse cultural norms.

Using Bayesian principles rather than a classical 'null hypothesis' approach, it is evident from the study's findings that the key to overcoming obstacles presented by human factors and economic pressures that might otherwise impede the development of a true safety culture in an organisation, is four-fold:

- Once they are identified those constraints and pressures resulting from organisational and cultural norms, legal and moral obligations, and economic considerations that impact upon organisational safety management, can be suitably dealt with by good strategic and organisational management using standard management techniques which in a culturally heterogeneous organisation require in addition the application of good cross-cultural management skills.
- Training can be used effectively to develop a common understanding of safety and safety systems, not only in individual companies but across the entire shipping industry. However, safety training alone is not sufficient to develop a true safety culture and reduce accidents: that requires additional emphasis on vocational and professional training.
- A company's training policies cannot be considered in isolation since they are to a large extent dictated by the company's employment philosophy.
- Uniform application of existing rules and regulations by all flag state administrations is a *sine qua non* for the establishment of a genuine safety culture throughout the world-wide shipping industry and is undoubtedly preferable to stricter enforcement of those rules and regulations on an arbitrary basis by port state inspectorates.

DISCUSSION

The book began by reviewing the need for and development of internationally agreed standards of maritime safety that led to the introduction of the ISM and STCW Codes in an attempt to ensure safety at sea, prevention of human injury or loss of life, and avoidance of damage to the environment and to property. Consideration was then given to the constraints and pressures that could prevent the Codes from achieving their goals, in particular those constraints and pressures that may be influenced by cultural values and attitudes and by psychological factors.

Such constraints and pressures exist at all levels of safety management, but from the model developed of the working of the ISM Code it was apparent that empirical research at levels 3, 4 and 5 of the safety management hierarchy would be the most appropriate to achieve the book's aims and objectives outlined above.

The empirical study highlighted the relevance and importance of effective education and training in raising safety standards. In particular:

- In Chapter 10 it was noted that of the 12 key informants interviewed:
    - eight had attended identical training courses covering ISM familiarisation and auditor training,
    - two had attended in-house ISM introductory seminars, and
    - two were qualified lead auditors of management systems.

    The training had resulted in harmonisation of the safety perspectives of all 12 persons.
- In Chapter 11 a connection was established between training and employment policies, and it was noted that all 12 key interviewees were agreed that raising the standards of education and training was the key to improving standards of safety throughout the shipping industry.
- Also in Chapter 11, computer training of sea-going personnel in Blue Ocean Offshore provided an example of how vocational training can be used to successfully overcome a specific safety-related problem.
- In Chapter 12 statistical analysis of seafarers' responses to the locus of control questionnaire provided some evidence, albeit tenuous, of a link between locus of control orientation and experience, thus lending some weight to the argument for greater emphasis on continuing education and training as a means of developing internal motivational cognition in emergency situations.

The existence of an obligation to develop personnel by means of suitable education and training is supported by decisions in a series of actions brought in English courts of law. In the case of *Donoghue (or McAlister) v Stevenson*[1] the judgement handed down by the House of Lords confirmed the existence of a duty of care that, in the words of Lord Atkin, requires you to '... take reasonable care to avoid acts or omissions which you can reasonably foresee would be likely to injure your neighbour', one's neighbour being identified as '... persons who are so closely and directly affected by my act that I ought reasonably to have them in contemplation as being so affected when I am directing my mind to the acts or omissions which are called in question'. The judgement was normative in so far as it not only confirmed the existence of a duty of care but also encapsulated in a few words the actions that ought to be taken in order to exercise that duty of care:

- identification of hazards;
- contemplation of risk;
- evaluation of the situation;
- determination of the steps to be taken to avoid damage to oneself, other people, property and the environment.

These actions are the very essence of modern safety management and the concept of a duty of care may therefore be looked upon as a basic norm in a hierarchy of norms from which safety management derives its authority.

Judicial decisions in subsequent cases built upon and developed the fundamental principle identified by Lord Atkin and further established that in consort with the duty of care there exists also a duty to keep abreast of developing knowledge and developing technology through continuing education and training, a duty that was succinctly summarised by Justice Swanwick in the case of *Stokes v Guest, Keen & Nettlefold (Bolts And Nuts) Ltd*[2] when he concluded that the overall test as to whether a duty of care had been properly exercised was undoubtedly the conduct of the reasonable and prudent employer taking positive thought for the safety of his or her workers in the light of what he or she knows or ought to know, and:

- where there is a recognised and general practice which has been followed for a substantial period in similar circumstances without mishap, he or she is entitled to follow it,

211

unless in the light of common sense or newer knowledge it is clearly bad;

- where there is developing knowledge, he or she must keep reasonably abreast of it and not be too slow to apply it; and
- where he or she has in fact greater than average knowledge of the risks, he or she may be thereby obliged to take more than average or standard precautions.

The principles articulated by Justice Swanwick were later adopted in the case of *Thompson v Smiths Shiprepairers (North Shields) (1984)*[3] by Justice Mustill who stated that common practice in an industry is relevant but not conclusive on the issue of negligence, not only when the negligence is said to be constituted by a failure to take known precautions, 'but also where the omission involves an absence of initiative in seeking out knowledge of facts which are not in themselves obvious'.

The decisions in these cases have done much to identify and clarify the duty both of organisations and of individuals:

- to act in a safe manner and with an acceptable and predictable level of both competence and competency; and
- to not only hire competent staff and competent contractors but also to identify training needs and ensure that staff receive continuing developmental instruction in order to meet changing organisational needs and advances in technology.

Justice Swanwick's summation of the general principles to be employed when deciding whether or not a duty of care has been properly exercised, supported by the statement of Justice Mustill, indicate the need for continuing instruction for both managers and workers in order that they might understand the risks associated with existing practices, keep abreast of risks in light of developing knowledge, and acquire the ability to raise the level of operational safety by the application of the highest standards of safety management.

Safety management, practical skills and theoretical knowledge are closely associated and the requirement for continuing instruction is now included in the ISM Code itself, the two most relevant clauses of the Code in this respect being:

Cl.1.2.2.3 Safety management objectives of the Company should, *inter alia:* continuously improve safety management skills of personnel ashore and aboard ships, including preparing for emergencies related both to safety and environmental protection.

Cl.6.5 The Company should establish and maintain procedures for identifying any training which may be required in support of the safety management system and ensure that such training is provided for all personnel concerned.

Shipping companies that interpret Clause 6 as merely a legal requirement to ensure that the seafarers it employs are properly qualified and in possession of the requisite certificates may have fulfilled their legal obligations. It is arguable however, whether they have fulfilled their moral obligation to interpret the clause not only to the letter but within the spirit of the Code as argued in Part I of the book, particularly in Chapters 2, 5 and 6.

Regulations and standards are by definition minimum requirements and must at the very least be complied with. As evidenced in the case of the M/V *Eurasian Dream*,[4] if those minimum standards are not complied with then a ship may be rendered unseaworthy. In that particular case the judge did indeed find the ship to be unseaworthy and held that 'Lack of due diligence is negligence and in this case there were numerous failures and errors of judgement that amounted to professional negligence in respect of the provision of equipment, competent master and crew and adequate documentation'.

But the spirit of the ISM Code is to introduce a genuine safety culture within the shipping industry and this implies training staff beyond the minimum stipulated requirements. Education and training provide the requisite knowledge and skills to undertake particular tasks while experience guides the individual in the application of that knowledge and those skills. Without sufficient skills and experience individuals may be unable to correctly tackle tasks that should be within their capability. Simply to try and plug gaps in knowledge or skills on an industry-wide basis by safety training alone is unlikely to achieve positive results.

If when developing their corporate training programmes senior managers observe only the minimum requirements of the STCW and

ISM Codes, and also misguidedly substitute safety training for vocational training and professional development, that may cause the decision making of employees to be influenced by cognitive biases based upon outdated information or false premises, possibly leading to unsound judgements and subsequent unsafe actions or inactions.

In order to make a sound decision, to be able to balance risk against reward, individuals must be in possession of appropriate knowledge, requisite skills and sufficient experience to be able to evaluate a situation, understand the consequences of their actions or inactions, and subsequently exercise suitable judgement. Without the knowledge, skills and experience to undertake a particular task, potential hazards associated with that task may not be identified or the correct course of action may not be apparent to the individual concerned and the consequences of a wrong decision may be far reaching.

For shipping companies to operate safely, their employees must be suitably educated, adequately trained and given sufficient opportunity to gain experience. This requires companies to ensure that people they employ are competent people and that continuing education and training are undertaken to keep them abreast of developments within their areas of expertise. It also requires the companies to provide employees with the opportunity to exercise newly acquired knowledge and skills in order to gain experience, for instance by developmental assignments.

However, although localised decisions about providing training to address failures in operational safety can be made at level 3 of the safety management hierarchy, it is clear from the ISM Code model (Chapter 8, Figure 9) that international cooperation at levels 1 and 2 of the safety management hierarchy is required to ensure even-handed application and enforcement of the provisions of both the ISM Code and the STCW Code in order to achieve an industry-wide safety culture and subsequent improvement in overall safety standards throughout the shipping industry. But as noted in Chapter 5, there has historically been a large disparity between flag states with regard to their stipulated minimum standards of education and training.

The implementation by IMO of the 1995 revision of the 1978 STCW Convention, which entered fully into force in 2002 and contains provisions for an internationally agreed basic minimum standard of training for the award of certificates of competency to sea-going personnel, may go some way to correcting that disparity. But a competence-based approach that simply provides an individual with basic training and the correct tools for the job would be insufficient. It is necessary also to teach people the underlying principles behind the work that they are expected to do and how the tools may be used to the best effect. Individuals must be provided with continuing professional education, vocational training and the opportunity to gain experience[5] otherwise they will not only be unable to carry out their work effectively, but may also become a liability to the safety of themselves, others, company property and the environment.

The following example of a series of mishaps that occurred on an offshore supply vessel operated by Blue Ocean Offshore will help to clarify the difference between competence-based training and a personal development approach, highlighting the importance of the latter with regard to improving safety standards.

The vessel was equipped with three diesel generators, any one of which was capable of supplying sufficient electrical power while the vessel was in port or on passage at sea. As a safety precaution and to allow for additional electrical load – for example when the bow thruster propeller was to be used for increased manoeuvrability – company procedures required a second generator also to be on line whenever the vessel was operating in close waters or manoeuvring within 500 metres of an offshore structure.

During a normal sea passage the watch-keeping oiler found the fluid flywheel of No.1 generator engine lying in the bilge. The function of a fluid flywheel is to dampen the vibrations of an internal combustion engine and without it the No.1 generator engine was operating with greater vibration than normal. The oiler retrieved the fluid flywheel and passed it to the Chief Engineer who unfortunately neither recognised what it was nor understood the ramifications of an internal combustion engine vibrating excessively. The engine continued to be routinely operated even though the vibration of the unit was increasingly greater than normal.

The Chief Engineer subsequently completed his tour of duty and was relieved by another Chief Engineer who not only recognised what the fluid flywheel was but also understood its function and the theory behind its application. He refitted it to the engine but unfortunately the vibration of the engine continued to be excessive.

He therefore reported the problem to his operations office and was instructed by the Fleet Manager not to use the engine until it had been inspected by the Fleet Technical Superintendent when the vessel returned to its home port.

Before the vessel returned to its home port, however, the governor of No.3 generator engine malfunctioned, restricting that generator to 30% of its rated output. The Chief Engineer was unable to identify the cause of the problem and was therefore instructed by the Fleet Manager to take the governor from the defective No.1 engine and fit it to the No.3 engine in order to maintain the vessel in a safe operating condition. Unfortunately, the Chief Engineer replied that he did not feel he had the requisite knowledge and skills to exchange the governors.

The vessel was therefore taken off hire by charterers and handed back to the owners for repair. Upon inspection it was found that:

- due to excessive vibration caused by running the engine without a fluid flywheel, No.1 generator engine was damaged beyond economical repair and a new engine had to be sourced, purchased and installed;
- the restricted output of No.3 generator was due simply to a bent rod linking the engine governor to the fuel rack. The link rod had been accidentally damaged and the incident had been neither reported nor discovered. Simply straightening the link rod rectified this deficiency.

In the above incidents both chief engineers, despite having the requisite qualifications for the positions they held, displayed an inability to apply fundamental engineering principles to situations that they had not previously encountered.

No amount of safety training would have increased the ability of the chief engineers to determine and apply to the situations they encountered the engineering principles involved, evaluate the problems correctly, exercise the right judgement and take the correct remedial action.

It could be argued that the two individuals concerned lacked specific task-related competences that affected their overall role competency. However, to provide every chief engineer throughout the shipping industry with the task-related competences necessary to deal with every conceivable emergency situation is unrealistic.

However, corporate training and development programmes designed to identify the strengths and weakness of individuals, and to provide continuing professional development, would help to develop such individuals not only by equipping them with the necessary knowledge and skills to carry out their work effectively but also by developing their ability to use deductive reasoning to solve problems that may arise which they have not previously encountered.

Chomsky[6] noted that in a totalitarian society the executive need not worry about what people think because the executive controls what people do, but in order to control what people do in a democracy the executive needs to control what people think. Applying this argument to raising safety standards throughout a world-wide, fragmented industry it is clearly not possible to control everything people do on board a ship or offshore structure, and therefore it is essential to ensure that they have the requisite knowledge, skills, training and experience to be able to think and act safely, to recognise hazards, take appropriate action to prevent hazards eventuating, to do their work effectively and hence to operate safely.

This cannot be achieved by safety training alone or simply by imparting competences. It can, however, be achieved by suitable vocational training together with continuing professional development of which safety training forms an integral part, and then providing individuals with developmental assignments to gain experience in accordance with their needs. But to ensure that such education and training are actually being provided and that they are of a suitable standard, it is necessary to audit the organisations responsible for its provision. To this end, the establishment of an international accreditation body, either as a singular body or as a network of institutions, is considered essential.

The authority to do this already rests with IMO because, as noted by Lord Donaldson,[7] an important administrative difference exists between the ISM Convention and the STCW Convention. Flag states signatory to the ISM Convention operate within a self-regulatory system under which the flag states certify to IMO that they are fully discharging their obligations under the Convention. The STCW Convention, on the other hand, contains a provision under which the signatory states delegate to IMO the authority to assess whether or not a signatory state is complying with its obligations under the Convention. How rigorously and with what ardour IMO monitor and enforce the provisions of the revised STCW Convention may well decide the effectiveness of the Convention in practice.

ENVOI

Previous enquiries into the introduction of the ISM Code have tended to concentrate upon the change elements involved and statistical surveys to determine whether the Code is actually achieving its objectives. In this book a completely different and primarily qualitative approach has been adopted that provided a richness of detail, analysis of which:

- determined that in order to improve industry-wide maritime safety standards, more emphasis on education and training would be more productive than concentrating on stricter enforcement of existing regulations or the introduction of further regulations;
- demonstrated that individual shipping companies not only can improve their safety standards by providing for the educational and training needs of their employees but have a legal duty and moral obligation to do so;
- provided a model for determining a company's safety culture maturity relative to previous audits or in comparison with a benchmark standard.

There were, however, two areas of limitation occasioned by the finite nature of the resources available to the author for research. First, while the book explored the impact of culturally influenced constraints and pressures at levels 3, 4 and 5 of the safety hierarchy because that is where safety management systems are developed and implemented, it is evident from the model of the ISM Code developed in Chapter 8 that culturally influenced constraints and pressures also operate at levels 1 and 2 of the safety hierarchy, and it is those constraints and pressures that influence the rigour with which nation states ensure that the provisions of the ISM Code are implemented by companies entering ships on their registers, and that education and training of a suitable standard is provided for their seafarers. However, empirical investigation of the effects of the constraints and pressures operating at levels 1 and 2 of the safety hierarchy was beyond the scope of the empirical study presented in this book.

Second, while the literature review indicated that similar constraints and pressures operate at levels 3, 4 and 5 of the safety hierarchy throughout the shipping industry, only one sector of the industry was examined in the empirical study. To demonstrate industry-wide commonality of the constraints and pressures and the manner in which they are handled, research encompassing other sectors of the shipping industry would be necessary.

However, although these areas of limitation provide opportunities for further research, the author is confident that they do not affect either the validity or the general applicability of the findings presented in this book.

Finished With Engines

Bibliography and Literature References

CHAPTER 1

1 Donaldson, Lord (2001), *A Rocky Road to Maritime Safety*. Paper presented at inaugural RNLI Annual Lecture, University of Southampton, School of Engineering Sciences, Ship Science Report No. 121.

2 Slater, P. (2001), 'Slow growth to no growth', *Lloyd's List*, 19th October 2001, p. 7.

3 Papalexis, E. (2001), *Tidal Wave in Shipping*. Unpublished paper presented at the Propeller Club, London, June 2001.

4 SSMR (2005), *World Shipbuilding and Maritime Casualties*, August/ September 2005. WWW http://www.Isl.org/products_services/publications/ pdf/COMM/8-9-2005-short.pdf- Accessed January 2006.

5 Amin, A. (1997) *Theory, Culture & Society* vol. 14(2): 123–137. SAGE, London, Thousand Oaks and New Delhi.

6 Özçayir, Z.O. (2001), *Port State Control*. Informa Professional, Informa Publishing Group Ltd., 69–77 Paul Street, London.

7 Talbot-Booth, E.C. (*circa* 1940), *His Majesty's Merchant Navy*. Sampson, Low, Marston & Co. Ltd., London, p.11.
Aldcroft, D.H. (1968), 'The Mercantile Marine', in D.H. Aldcroft, ed., *The Development of British Industry and Foreign Competition 1875–1914; Studies in Industrial Enterprise*. Toronto, University of Toronto Press.
Podolny, J.M., & Scott-Morton, F.M. (1998), *Social Status, Entry and Predation: The Case of British Shipping Cartels 1879–1929*. Stanford University GSB, Stanford, CA 94305–5015, USA.

8 *Lloyd's Maritime Directory 2001* (2000). Informa Publishing Group Ltd., Colchester, Essex.

9 Guide to Shipbuilding, Repair and Maintenance 2000/2001 (2000), *Lloyd's Ship Manager*. Colchester, Essex.

10 Seafarers International Research Centre (1999), 'The Global Market for Seafarers'. *Proceedings of SIRC's First Symposium at Cardiff University*, Cardiff, UK.

11 International Transport Workers Federation (2003), *Flags of Convenience*, ITF Website: WWW http://www.itfglobal.org/flags-convenience/index. cfm Accessed January 2006.

12 Fröbel, Folker, Heinrichs, Jürgen, and Kreye, Otto (1980), *The new*

international division of labour. Structural unemployment in industrialized countries and industrialization in developing countries. Cambridge University Press, Cambridge, UK.

13 Organisation for Economic Cooperation and Development (1996), *Competitive Advantages Obtained by Some Ship owners as a Result of Non-observance of International Rules and Standards.* OECD, Paris, France.

14 Health and Safety Executive (1992), *Management of Health and Safety Regulations,* SI 1992/2051.

15 Beale, S. (2004), 'Risk Rates Rocket As Shipyard Accidents Increase', report in the Institute of Marine Engineering Science and Technology electronic journal *emarine, issue 21:* WWW http://www.imarest.org/emarine/issue21.htm Accessed 30th November 2004.

16 Drucker, P.F. (1996) *The Practice of Management.* First Published by Butterworth-Heinemann 1955, Re-Published 1996.

17 Weyman, A. (1998), *Risk Perception and Risk Taking Behaviour at Work: A Review of Literature.* Health & Safety Executive, Sheffield.

18 Lau, C.M., McMahan, G.C., Woodman, R.W. (1996), 'An International Comparison of Organization Development Practices', *Journal of Organizational Change Management,* vol.9 No.2, pp 4–19, MCB University Press.

19 McMahan, G.C., and Woodman, R.W. (1992), 'The Current Practice of Organization Development Within the Firm', *Group and Organization Management,* vol.17 no.2, pp 117–34.

20 Barr, S. (1999), 'ISM Set to Bring Benefit of Lower Claims: Marine Cover'. *Lloyd's List,* 21st May 1999. Informa Publishing Group Ltd., Colchester, Essex.

21 Mulrenan, J. (2003), 'Owners Split over Value of ISM Code', *Trade Winds,* 12th September 2003, p.19. Trond Lillestolen, Oslo, Norway.

22 Anderson, P. (2002), *Managing Safety at Sea.* Doctoral Thesis, Middlesex University, November 2002.

23 De Bievre, A. (2001), 'Pitfalls of the new ISM Code', *Lloyd's List,* 19 Feb. 2001, p. 20. Informa Publishing Group Ltd., Colchester, Essex.

24 IMO (2007), WWW http://www.imo.org/Conventions/contents.asp?doc_id=651&topic_id=257#13 Accessed 30th May 2007.

25 Trompenaars, F., and Hampden-Turner, C. (2000), *Riding The Waves Of Culture,* Second Edition. Nicholas Brealey Publishing, London.

26 Geertz, C. (1973), *The Interpretation of Cultures.* Basic Books, New York.

27 Schein, E. (1992), *Organisational Culture and Leadership,* 2nd Edition. Jossey-Bass, San Francisco, California, USA.

28 Hofstede, G. (1991), *Cultures and Organisations.* HarperCollins Business, Hammersmith, London.

29 Holden, N.J. (2002), *'Cross-cultural Management: A Knowledge Management Perspective'.* Pearson Education Ltd, Harlow, England.

30 ACSNI Human Factors Study Group, Advisory Committee on the Safety of Nuclear Installations (1993), *Third Report: Organising for Safety.* HMSO, London, ISBN 0118821040.

31 Cooper, M.D. (2000), 'Towards a Model of Safety Culture', *Safety*

Science, Vol 36, pp 111–136. Journal published by Applied Behavioural Sciences Ltd., Hull, East Yorkshire.

32 Young, E. (1989), 'On the Naming of the Rose: Interests and Multiple Meanings as Elements of Organisational Culture, *Organization Studies*, 10/2: 187–206. Manchester Business School, Manchester, UK.

33 Spybey, T. (1996), 'Global Transformations' in *'Translating Organisational Change'*, edited by Barbara Czarniawska and Guje Sevón. Walter de Gruyter, Berlin and New York.

34 Czarniawska, B., and Joerges, B. (1996), 'Travels of Ideas', published in the book *Translating Organizational Change*, edited by Czarniawska, B., and Sevón, G., Walter de Gruyter, Berlin and New York.

35 Hale, A.R. (1984), 'Is safety training worthwhile?', *Journal of Occupational Accidents* 6: 17–33. ISSN: 0376–6349.

36 Rotter, J.B. (1966), 'Generalized expectancies for internal versus external control of reinforcement'. *Psychological Monographs* 1966, 80 (whole no. 609).

CHAPTER 2

1 IMO (1995), *International Convention on Standards of Training, Certification and Watchkeeping for Seafarers 1978, Seafarers' Training, Certification and Watchkeeping (STCW) Code*, STCW.6/Circ.1, Ref.A1/V/3.02. International Maritime Organization, 4 Albert Embankment, London, 24 July 1995.

2 IMO (2005), *Status of Multilateral Conventions and Instruments in Respect of Which the International Maritime Organisation or its Secretary-General Performs Depository or Other Functions as at 31 December 2005*, Resolution MSC.99(7.3), I:\J_\9193.doc. International Maritime Organization, 4 Albert Embankment, London.

3 Sagen, A. (1999), *The ISM Code In Practice*. Tano Aschehoug, Norway.

4 Shaw, A. (2001), 'Desperately Seeking A Safety Culture', *The Safety & Health Practitioner*, March 2001, pp. 21–22.

5 Carnall, C. 1995 *Managing Change in Organisations*, Second Edition. Prentice Hall International (UK) Ltd. p.120.

6 Hale, A.R. (2000), 'Culture's confusions', *Safety Science* vol.34, no.1–3, pp.1–14.

Hofstede, G. (1991), *Cultures and Organisations*. Harper Collins Business, Hammersmith, London.

ACSNI Human Factors Study Group, Advisory Committee on the Safety of Nuclear Installations (1993), *Third Report: Organising for Safety*. HMSO, London, ISBN 0118821040.

CHAPTER 3

1 MacLean, R. (1994), *Public International Law Textbook*, 16th edition. HLT Publications, London.
2 Özçayir, Z.O. (2001), *Port State Control*. Informa Professional, Informa Publishing Group Ltd., 69–77 Paul Street, London.
3 Churchill, R.R., and Lowe, A.V. (1999), *The Law of the Sea*, Third Edition. Melland Schill Studies in International Law, Juris Publishing Manchester University Press, Oxford Road, Manchester.
   Özçayir, Z.O. (2001), *Port State Control*. Informa Professional, Informa Publishing Group Ltd., 69–77 Paul Street, London.
4 Nottebohm Case: Liechtenstein v Guatemala (1955), ICJ Rep., p.4.
5 International Transport Workers Federation (2003), *Flags of Convenience*. ITF Website: WWW http://www.itfglobal.org/flags-convenience/index. cfm Accessed January 2006
6 Africa Confidential (2005), Vol. 46 No.10, 13th May 2005, quoting extract from Vol. 42 No.22, 9th November 2001 WWW www.africa-confidential.com/index.aspx?pageid = 22&countryid = 27 Accessed 15th May 2005.
7 *see for example:*
   Office of the Maritime Administrator, Marshall Islands Maritime and Corporate Administrators Inc., 437 Madison Avenue, New York, USA.
   Liberian International Ship & Corporate Registry, Ship Registration & Documentation, 99 Park Avenue, Suite 1700, New York, NY 10016–1601, USA.
   The Bahamas Maritime Authority, 120 Old Broad Street, London EC2N 1AR, UK.
   St Kitts & Nevis International Ship Registry, West Wing, York House, 48–50 Western Road, Romford, Essex RM1 3LP, UK.
8 Asylum Case: Columbia v Peru (1950), ICJ Rep p.266.
9 North Sea Continental Shelf Cases: Federal Republic of Germany v Denmark; Federal Republic of Germany v The Netherlands (1969), ICJ Rep p.14.
10 Donaldson, Lord (2001), *A Rocky Road to Maritime Safety*. Paper presented at inaugural RNLI Annual Lecture, University of Southampton, School of Engineering Sciences, Ship Science Report No. 121.
11 Wildenhus's Case (1887), United States Supreme Court, Wait CJ, 120 US 1.
12 Nimbus case: Sellers v Maritime Safety Inspector (1998), New Zealand Court of Appeal; CA104/98, New Zealand Law Reports, [1999] 2 NZLR 44; 1998 NZLR LEXIS 650.

CHAPTER 4

1   Le Guen, J. (1999), *Reducing Risks, Protecting People*. Discussion Document, Risk Assessment Policy Unit, Health & Safety Executive, HSE Books, Sudbury Suffolk.

2   Robens Committee, Committee on Safety and Health at Work (1972), *Safety and Health at Work*. HMSO, ISBN 0 10 150340 7.

3   Kelsen, H. (1967), *Pure Theory of Law*, trans. Max Knight. University of California Press, Berkeley and Los Angeles.

4   Wacks, R. (1993), *Jurisprudence*, SWOT Series, 3rd edition. Blackstone Press Ltd., London.

5   Harris, J.W. (2004), *Legal Philosophies*, (Second Edition), first published 1997, rep. 2004. Oxford University Press, Great Clarendon Street, Oxford.

6   Williams, J.L. (1960), *Accidents and Ill Health at Work*. Staple Press, London.

7   ACSNI Human Factors Study Group, Advisory Committee on the Safety of Nuclear Installations (1993), *Third Report: Organising for Safety*. HMSO, London, ISBN 0118821040.

8   Watson, S.R. (1981), 'Risks and Acceptability'. *Journal of the Society of Radiological Protection*, vol.1(4): pp.21–25.

9   Pidgeon, N., Hood, C., Jones, D., Turner, B., & Gibson, R. (1992), *Risk Analysis, Perception and Management*. The Royal Society, London.

10 Royal Society Report (1992), *Risk analysis, perception and management* (1992).

11 Sagen, A. (1999), *The ISM Code In Practice*. Tano Aschehoug, Norway.

12 Shaw, A. (2001), 'Desperately Seeking A Safety Culture', *The Safety & Health Practitioner*, March 2001, pp. 21–22.

13 Cooper, M.D. (2000), 'Towards a Model of Safety Culture', *Safety Science*, Vol 36, pp 111–136. Journal published by Applied Behavioural Sciences Ltd., Hull, East Yorkshire.

14 Young, E. (1989), 'On the Naming of the Rose: Interests and Multiple Meanings as Elements of Organisational Culture, *Organization Studies*, 10/2: 187–206. Manchester Business School, Manchester, UK.

15 Carnall, C. (1995), *Managing Change in Organisations*, second edition. Prentice Hall International (UK) Ltd. p.120.

16 Hale, A.R. (1984), 'Is safety training worthwhile?', *Journal of Occupational Accidents* 6: 17–33. ISSN: 0376–6349.
    Hofstede, G. (1991), *Cultures and Organisations*. HarperCollins Business, Hammersmith, London.
    ACSNI Human Factors Study Group, Advisory Committee on the Safety of Nuclear Installations (1993), *Third Report: Organising for Safety*. HMSO, London, ISBN 0118821040.

17 Bandura, A. (1986), *Social Foundations of Thought and Action: A Social Cognitive Theory*. Prentice-Hall, Englewood Cliffs, NJ.

18 Genn, H. (1987), *Great Expectations: the Robens legacy and employer self-*

regulation. Unpublished paper presented to the Health & Safety Executive, cited in 1993 report by ACSI Human Factors Study Group.

19 United Kingdom Health and Safety Executive (1992), *Management of Health and Safety Regulations*, SI 1992/2051.

20 ACSNI Human Factors Study Group, Advisory Committee on the Safety of Nuclear Installations (1993), *Third Report: Organising for Safety.* HMSO, London, ISBN 0118821040.

Le Guen, J. (1999), *Reducing Risks, Protecting People.* Discussion Document, Risk Assessment Policy Unit, Health & Safety Executive, HSE Books, Sudbury Suffolk.

Harvey, J., Bolam, H., Gregory, D., Erdos, G. (2001), 'The effectiveness of training to change safety culture and attitudes within a highly regulated environment', *Personnel Review*, vol 30, no.5–6, 2001, pp. 615–636.

21 Argyris, C. (1977), *Double-loop learning in organizations.* Harvard Business Review, 55 (5): 115–25.

22 Argyris, C., and Schön, D.A. (1978), *Organizational Learning.* Addison-Wesley, Reading, MA.

23 Jankowicz, A.D. (2000), *Business Research Projects*, third edition. Business Press, Thomson Learning, London.

24 Dupont Safety Resources (1999), *The DuPont safety Philosophy.* E.I. du Pont de Nemours and Company, USA.

25 Fleming, M. (2001), *Safety Culture Maturity Model.* Health & Safety Executive, HSE Books, Sudbury, Suffolk, ISBN 0 7176 1919 2.

26 Donaldson, Lord (2001), *A Rocky Road to Maritime Safety.* Paper presented at inaugural RNLI Annual Lecture, University of Southampton, School of Engineering Sciences, Ship Science Report No. 121.

27 Holden, N.J. (2002), '*Cross-cultural Management: A Knowledge Management Perspective*'. Pearson Education Ltd, Harlow, England.

28 Hofstede, G. (1991), *Cultures and Organisations.* HarperCollins Business, Hammersmith, London.

CHAPTER 5

1 Gadd, S., and Collins, A.M. (2002), *Safety Culture: A Review of the Literature.* HSL/2002/25, Health & Safety Laboratory, Sheffield.

Schmidt, R., Lyytinen, K., Keil, M., and Cule, P. (2001), 'Identifying Software Project Risks: An International Delphi Study', *Journal of Management Information Systems*, Spring 2001, Vol. 17, No. 4, pp. 5–36.

Krampen, G. and Weiberg, H. (1981), Three aspects of locus of control in German, American, and Japanese university students. *Journal of Social Psychology*, 113, 133–134.

2 Tayeb, M.H. (1995), 'The Competitive Advantage of Nations: The Role of HRM and its Socio-cultural Context', *International Journal of Human Resource Management*, vol.6, pp. 588–605.

3   Pheysey, D.C. (1993), *Organisational Cultures*. Routledge, New York, NY 1001.
4   Schein, E. (1992), *Organisational Culture and Leadership*, 2nd Edition. Jossey-Bass San Francisco, California.
5   Shaw, A. (2001), 'Desperately Seeking A Safety Culture', *The Safety & Health Practitioner*, March 2001, pp. 21–22.
6   Glendon, A.I., and McKenna, E.F. (1995), *Human safety and Risk Management*. Chapman and Hall, London.
     Confederation of British Industry (CBI) (1990), *Developing a Safety Culture – Business for safety*. CBI, London.
     Hale, A.R. (2000), 'Culture's confusions', *Safety Science* vol.34, no.1–3, 1–14.
7   Lee, T., and Harrison, K. (2000), 'Assessing Safety Culture at a Nuclear Reprocessing Plant', *Safety Science*, 30, pp.61–97.
8   Carnall, C. (1995), *Managing Change in Organisations*, second edition. Prentice Hall International (UK) Ltd. p.120.
     Pheysey, D.C. (1993), *Organisational Cultures*. Routledge, New York, NY1001.
9   Hale, A.R. (1984), 'Is safety training worthwhile?', *Journal of Occupational Accidents* 6: 17–33. ISSN: 0376–6349.
     Hofstede, G. (1991), *Cultures and Organisations*. HarperCollins Business, Hammersmith, London.
     ACSNI Human Factors Study Group, Advisory Committee on the Safety of Nuclear Installations (1993), *Third Report: Organising for Safety*. HMSO, London, ISBN 0118821040.
10 Holden, N.J. (2002), *'Cross-cultural Management: A Knowledge Management Perspective'*. Pearson Education Ltd, Harlow, England.
11 Trompenaars, F., and Hampden-Turner, C. (2000), *Riding The Waves Of Culture*, second edition. Nicholas Brealey Publishing, London.
12 Inkeles, A., and Levinson, D.J. (1969), 'National character; the study of modal personality and sociocultural systems', in *The Handbook of Social Psychology*, 2nd edn., vol.4, G. Lindsay & E. Aronson (eds). Reading MA: Addison-Wesley.
13 Hofstede, G., and Bond, M.H. (1988), 'The Confucius connection: from cultural roots to economic growth', *Organisational Dynamics*, vol.16, issue 4, pp. 4–21.
14 Heider, F. (1958), *The Psychology of Interpersonal Relations*. New York: Wiley.
15 Rotter, J.B. (1966), 'Generalized expectancies for internal versus external control of reinforcement', *Psychological Monographs* 1966, 80 (whole no. 609).
     Kelly, H.H. (1967), 'Attribution Theory in Social Psychology', *Nebraska Symposium on Motivation,* vol.15, Edited by D. Levine and M. Lincoln. University of Nebraska Press.
     Weiner, B. (1974), *Achievement, motivation and attribution theory*. General Learning Press, Morristown, NJ, USA.
     Calder, B. (1977), 'An Attribution Theory of Leadership', in Staw, B., and

Salanick, G., (Eds), *New Directions in Organizational Behaviour, pp. 179–204*. St. Clair Press, Chicago, IL, USA.

16 Barnett, C.K. (1999), WWW http://pubpages.unh.edu/~ckb/attribution-figures.html Accessed 16th July 2004.

17 Simons, Irwin and Drinnin (1987), *Psychology: The Search for Understanding*. West Publishing.
    Trompenaars, F., and Hampden-Turner, C. (2000), *Riding The Waves Of Culture*, second edition. Nicholas Brealey Publishing, London.

18 Mintzberg, H. (1979), *The Structure of Organisations*. Prentice-Hall, Englewood Cliffs, New Jersey.

19 Mintzberg, H., and Waters, J.A. (1985), Of Strategies: Deliberate and Emergent, Strategic *Management Journal*, July/September, 1985, pp. 257–272. John Wiley and Sons Ltd.

20 Gaa, J. and Shores, J. (1979), Domain specific locus of control among Black, Anglo, and Chicano undergraduates. *Journal of Social Psychology*, 107, 3–8.
    Krampen, G. and Weiberg, H. (1981), Three aspects of locus of control in German, American, and Japanese university students. *Journal of Social Psychology*, 113, 133–134.

21 Bayne, G. (2002), Unpublished research, unpublished dissertation: *The Relationship Between Individualism/Collectivism, Locus of Control And Sense of Coherence*, WWW http://www.usq.edu.au/users/bayneg/research.htm Accessed 18th May 2004.

22 Furnham, A. and Henry, J. (1980), Cross-cultural locus of control studies: experiment and critique. *Psychological Reports*, 47, 23–29.

23 Haley, U.C.V., and Stumpf, S.A. (1989), 'Cognitive trials in strategic decision-making: linking theories of personalities and decisions', *Journal of Management Studies*, 26,5, 477–497.

24 Kahneman, D., Slovik, P. and Tversky A. (1982), *Judgement under Uncertainty: Heuristics and Biases*. Cambridge University Press.

25 Rundmo, T. (2000), 'Safety Climate, Attitudes and Risk Perception in Norsk Hydro', *Safety Science*, vol.34, no.103, pp.47–59.

26 Heur, R.J. Jr. (1999), *Psychology of Intelligence Analysis*. Centre for the Study of Intelligence, Washington, DC.

27 Gadd, S., and Collins, A.M. (2002), *Safety Culture: A Review of the Literature*, HSL/2002/25. Health & Safety Laboratory, Sheffield.

28 Le Guen, J. (1999), *Reducing Risks, Protecting People*. Discussion Document, Risk Assessment Policy Unit, Health & Safety Executive, HSE Books, Sudbury Suffolk.

29 Douglas, M.S., and Wildavsky, A. (1982), *Risk and Culture*. University of California Press, Berkeley.
    Funtowicz, S.O., and Ravetz, J.R. (1992), 'Three Types of Risk Assessment and the Emergence of Post-normal Science', in *Social Theories of Risk*, ed. Kirmsky, S., and Golding, D., Praeger. Westport, Connecticut, pp.251–274.
    Pidgeon, N., Hood, C., Jones, D., Turner, B., & Gibson, R. (1992), *Risk Analysis, Perception and Management*. The Royal Society, London.

227

30 Hisrich, R.D., Peters, M.P., and Shepherd, D.A. (2005), *Entrepreneurship*, 6th Edition, International Edition. McGraw-Hill/Irwin, New York, USA.

31 Nonprofit Risk Management Center (2001), Washington, DC, USA, WWW, http://www.nonprofitrisk.org/whatis.htm Accessed 6 April 2001.

32 Pidgeon, N., and O'Leary, M. (2000), 'Man-made disasters: why technology and organisations (sometimes) fail', *Safety Science*, vol.34, pp.15–30.
Cox, S., and Flin, R. (1998), 'Safety Culture: philosopher's stone or man of straw?', *Work and Stress*, No.12, vol.3, pp.189–201.
Cheyne, A., Cox, S., Oliver, A., and Thomas, J.M. (1998), 'Modelling safety climate in the prediction of levels of safety activity', *Work and Stress*, vol.12(3), pp.255–271.

33 Etzioni, A. (1975), *A Comparative Analysis of Complex Organisations*. Free Press, New York.

34 McGill University (2002), *Impact of Hofstede's Cultural Dimensions on Management Issues*, WWW http://www.management.mcgill.ca/course/orgbeh/JAEGER/HOFSMGMT.HTM Accessed 20 January 2006.

35 Lave, J., and Wenger, E. (1991), *Situated Learning: Legitimate Peripheral Participation*. Cambridge University Press, New York.

36 Easterby-Smith, M., and Araujo, L. (1999), 'Organisational Learning: Current Debates and Opportunities', in '*Organizational Learning and the Learning Organization*', edited by Easterby-Smith, M., Burgoyne, J., and Araujo, L. Sage Publications, London, 1999.

37 Organisation for Economic Co-operation and Development (1996), *Competitive Advantages Obtained by Some Ship owners as a Result of Non-observance of International Rules and Standards*. OECD, Paris, France.
Squire, D. (2005), 'Competent people make the difference', *Alert: The international Maritime Human Element Bulletin*. The Nautical Institute, London.

38 Hale, A.R. (1984). 'Is safety training worthwhile?', *Journal of Occupational Accidents* 6: 17–33. ISSN: 0376-6349.

39 Taylor, F.W. (1911), *The Principles of Scientific Management*. Harper and Row, New York, USA.

40 Organisation For Economic Co-operation And Development (1996), *Competitive Advantages Obtained by Some Ship owners as a Result of Non-observance of International Rules and Standards*. OECD, Paris, France.
Donaldson, Lord (2001), *A Rocky Road to Maritime Safety*. Paper presented at inaugural RNLI Annual Lecture, University of Southampton, School of Engineering Sciences, Ship Science Report No. 121.

CHAPTER 6

1 Harris, J.W. (2004), *Legal Philosophies*, (Second Edition), first published 1997, rep. 2004. Oxford University Press, Great Clarendon Street, Oxford.

2 Trafford, S.M. (May 2005), *Towards a Deeper Understanding of Safety*.

Unpublished monograph, University of Luton, School of Business Studies, 1st May 2005.

3 MacLean, R. (1994), *Public International Law Textbook*, 16th edition. HLT Publications, London.

4 Furmston, M.P. (2001), *Cheshire, Fifoot & Frumston's Law of Contract*. Fourteenth Edition, October 2001. Butterworths, London.

5 Trompenaars, F., and Hampden-Turner, C. (2000), *Riding The Waves Of Culture*. Second Edition. Nicholas Brealey Publishing, London.

6 Ehrlich, E. (1936), 'Fundamental principles of the Sociology of law', transl. W.L. Moll. *Harvard Studies in Jurisprudence*, vol.5, Harvard University Press, Cambridge, Mass.

7 Wacks, R. (1993), *Jurisprudence*, SWOT Series, 3rd edition. Blackstone Press Ltd., London.

8 Macauley, S. (1963), 'Non-contractual Relations in Business, A Preliminary Study', *American Sociological Review*, vol.28, p.55 et seq.

9 Beale, H., and Dugdale, T. (1975), *Contracts Between Businessmen*, British Journal of Law and Society, vol.2, pp. 45–60.

10 Unger, R.M. (1977), *Law in Modern Society, Toward a Criticism of Social Theory*. Collier MacMillan, London, p.250.

11 Walker, R.J. (1980), *The English Legal System*. Butterworths, London. Furmston, M.P. (2001), *Cheshire, Fifoot & Frumston's Law of Contract*, Fourteenth Edition, October 2001. Butterworths, London.

12 Krysakowska-Budny E., and Jankowicz, A.D. (1991) 'Poland's road to capitalism', *Salisbury Review 1991*, 10, 1, 28–31.

13 Cooke, R. (1989), 'Fairness', *Victoria University Weekly Law Review*, pp. 421–423.

14 Papera Traders Co Ltd v Hyundai Merchant Marine Co Ltd (2002), *M/V Eurasian Dream*, Cresswell J, Lloyds Law Reports 2002, vol 1; Part 11, Lloyd's Rep 719 QB.

15 Davis v Stena Line Ltd (2005), Case No: HQ03X01321, High Court, QB Division, Royal Courts of Justice, The Strand, London, 17th March 2005 before Mr Justice Forbes.

16 Rawls, J. (1972), *A Theory of Justice*. Oxford University Press, London.

17 Kelsen, H. (1957), 'Why should the law be obeyed?' in *'What is Justice'? Justice, Law and Politics in the Mirror of Science*. University of California press, Berkeley and Los Angeles.

18 Donaldson, Lord (2001), *A Rocky Road to Maritime Safety*. Paper presented at inaugural RNLI Annual Lecture, University of Southampton, School of Engineering Sciences, Ship Science Report No. 121.

19 *See the following report:*
Organisation for Economic Co-operation and Development, (1996), *Competitive Advantages Obtained by Some Ship owners as a Result of Non-observance of International Rules and Standards*. OECD, Paris, France.

20 Shaw, A. (2001), 'Desperately Seeking A Safety Culture', *The Safety & Health Practitioner*, March 2001, pp. 21–22.

CHAPTER 7

1 Czarniawska, B., and Joerges, B. (1996), 'Travels of Ideas', published in the book *Translating Organizational Change*, edited by Czarniawska, B., and Sevón, G. Walter de Gruyter, Berlin and New York.

2 Rogers, E.M. (1962), *Diffusion of innovations*. Free Press, New York. Levitt, B., and March, J.G. (1988), 'Organizational Learning', *Annual Review of Sociology*, 14: 319–340.

3 Latour, B. (1986), *Power, action and belief*. Routledge and Kegan Paul, London, pp 264–280.

4 Amin, A. (1997), *Theory, Culture & Society*. SAGE, London, Thousand Oaks and New Delhi, vol. 14(2): 123–137.

5 Costea, B. (1999), *International MBAs and globalisation: celebration or end of diversity*. Lancaster University.

6 Mann, T. (1928), *The Magic Mountain*. Secker, London.

7 Spybey, T. (1996), 'Global Transformations' in '*Translating Organisational Change*', edited by Barbara Czarniawska and Guje Sevón. Walter de Gruyter, Berlin and New York.

8 Brealey, R.A., and Meyers, S.C. (1996), *Principles of Corporate Finance*. McGraw-Hill, USA.

9 Jankowicz, A.D. (1998), 'Planting a paradigm in central Europe: do we graft, or must we wet the rootstock anew?', *Management Learning*, 1998, 30, 3, 281–299.

10 Heidegger, M., quoted by Twomey, D.V. (2003), in *The End of Irish Catholicism*. Veritas Publications, Dublin, Ireland.

11 Hirst, P.Q., and Thompson, G. (1996), *Globalization in Question*. Polity Press, London.

12 *Daily Mail* – Thursday 20th July 2000 – 'Booming Britain's £164bn world takeover' – at page 2.

13 Holden, N.J. (2002), '*Cross-cultural Management: A Knowledge Management Perspective*'. Pearson Education Ltd, Harlow, England.

14 Hofstede, G. (1991), *Cultures and Organisations*. HarperCollins Business, Hammersmith, London.

15 Le Guen, J. (1999), *Reducing Risks, Protecting People*. Discussion Document, Risk Assessment Policy Unit, Health & Safety Executive, HSE Books, Sudbury Suffolk.

16 Drucker, P.F. (1996) *The Practice of Management*. First Published by Butterworth-Heinemann1955, Re-Published 1996.

17 Mobil Oil (Summer 1999), *Mobil Prospect*, an in-house magazine published by Mobil Oil Company, Public Affairs Department, Mobil Court, London.

18 ExxonMobil, February 2005, *NEWSline*, an in-house magazine published by ExxonMobil, London.

CHAPTER 8

1   Jankowicz, A.D. (2000), *Business Research Projects*, third Edition. Business Press, Thomson Learning, London.
2   Dominica Maritime Administration, advertisement in *Lloyd's List* (22 October 2001), Informa Publishing Group Ltd., Colchester, Essex.
3   Cambodia Ship Registry, advertisement in *Lloyd's List* (31 October 2001), Informa Publishing Group Ltd., Colchester, Essex.

CHAPTER 9

1   Trafford, S.M. (June 2006), *The Impact of the Diversity of Cultures Upon the Implementation of the International Management Code for the Safe Management of Ships and for Pollution Prevention.* A Doctoral Thesis submitted to the University of Bedfordshire, Luton, UK.
2   Yin, R. (1994), *Case Study Research: Design and Methods.* Sage, London.
3   Hofstede, G. (1991), *Cultures and Organisations.* HarperCollins Business, Hammersmith, London.
4   Rotter, J.B. (1966), 'Generalized expectancies for internal versus external control of reinforcement', *Psychological Monographs* 1966, 80 (whole no. 609).
5   *See for example:*
ACSNI Human Factors Study Group, Advisory Committee on the Safety of Nuclear Installations (1993), *Third Report: Organising for Safety.* HMSO, London, ISBN 0118821040.
6   Blaxter, L., Hughes, C., Tight, M. (1996), *How to Research.* Open University Press, Buckingham, Philadelphia, USA.
7   Tremblay, M.A. (1982), 'The key informant technique: a non-ethnographic application' in Burgess R. (ed.) *Field Research: a Sourcebook and Field Manual.* London: Allen and Unwin.

CHAPTER 10

1   Schein, E. (1992), *Organisational Culture and Leadership*, 2nd Edition. Jossey-Bass San Francisco, California, p.328.
2   Hofstede, G. (1991), *Cultures and Organisations.* HarperCollins Business, Hammersmith, London.
3   Tremblay, M.A. (1982), 'The key informant technique: a non-ethnographic application' in Burgess R. (ed.) *Field Research: a Sourcebook and Field Manual.* London: Allen and Unwin.
4   Pidgeon, N., and O'Leary, M. (2000), 'Man-made disasters: why technology and organisations (sometimes) fail', *Safety Science*, vol.34, pp.15–30.

Cox, S., and Flin, R. (1998), 'Safety Culture: philosopher's stone or man of straw?', *Work and Stress*, No.12, vol.3, pp.189–201.

Cheyne, A., Cox, S., Oliver, A., and Thomas, J.M. (1998), 'Modelling safety climate in the prediction of levels of safety activity', *Work and Stress*, vol.12(3), pp.255–271.

5   Etzioni, A. (1975), *A Comparative Analysis of Complex Organisations.* Free Press, New York.

CHAPTER 11

1   ACSNI Human Factors Study Group, Advisory Committee on the Safety of Nuclear Installations (1993), *Third Report: Organising for Safety.* HMSO, London, ISBN 0118821040.

2   OCIMF (1997), *Marine Injury Reporting Guidelines.* Oil Companies International Marine Forum, 27 Queen Anne's Gate, London.

3   International Support Vessel Owners' Association (2004), *ISOA Personnel Accident Survey 2003.* International Support Vessel Owners' Association, 12 Carthusian Street, London.

CHAPTER 12

1   Rotter, J.B. (1966), 'Generalized expectancies for internal versus external control of reinforcement', *Psychological Monographs* 1966, 80 (whole no. 609).

2   NUMAST (2004), Memorandum by NUMAST (TT01), June 2004, *Written Evidence To Select Committee On Transport*, The United Kingdom Parliament, WWW www.publications.parliament.uk/pa/cm200405/ cmselect/cmtran/299/299we02.htm Accessed 15th Dec 2005.

3   Hsu, J.C. (1996), '*Multiple Comparisons Theory and Methods*'. Chapman & Hall, London.

4   Hofstede, G. (1991), *Cultures and Organisations.* HarperCollins Business, Hammersmith, London.

CHAPTER 13

1   ACSNI Human Factors Study Group, Advisory Committee on the Safety of Nuclear Installations (1993), *Third Report: Organising for Safety.* HMSO, London, ISBN 0118821040.

2   Cooper, M.D. (2000), 'Towards a Model of Safety Culture', *Safety Science*, Vol 36, pp 111–136. Journal published by Applied Behavioural Sciences Ltd., Hull, East Yorkshire.

3   Personal communication, 15th July 2004.

4   Personal communication, 22nd August 2005.
5   OCIMF (2008), *Tanker Management and Self Assessment*, 2nd edition, Witherby Seamanship International, Livingston, UK.

CHAPTER 14

1   *Donoghue (or McAlister) v Stevenson*, AC 562, 1932, House of Lords.
2   *Stokes v Guest, Keen & Nettlefold(Bolts And Nuts) Ltd.*, Assizes (1968), 1 W.L.R. 1776; 112S.J. 821.
3   *Thompson v Smiths Shiprepairers (North Shields) Ltd* [1984] QB 405.
4   *Papera Traders Co Ltd v Hyundai Merchant Marine Co Ltd (2002)*, 'M/V Eurasian Dream', Cresswell J, Lloyds Law Reports 2002, Vol 1; Part 11, Lloyd's Rep 719 QB.
5   Munro-Faure, L., and Munro-Faure, M. (1992), *Implementing Total Quality Management. Financial Times*, Pitman Publishing, 128 Long Acre, London.
6   Chomsky, N., (1991), *Media Control*, WWW. http://www.zmag.org/chomsky/talks/9103-media-control.html Accessed 18th October 2005.
7   Donaldson, Lord (2001), *A Rocky Road to Maritime Safety*. Paper presented at inaugural RNLI Annual Lecture, University of Southampton, School of Engineering Sciences, Ship Science Report No. 121.